PERACETIC ACID DISINFECTION—IMPLEMENTATION CONSIDERATIONS FOR WATER RESOURCE RECOVERY FACILITIES

WEF Special Publication

First Edition

2020

Water Environment Federation
601 Wythe Street
Alexandria, VA 22314-1994 USA
https://www.wef.org

ISBN 978-1-57278-357-7

About WEF

The Water Environment Federation (WEF) is a not-for-profit technical and educational organization of 35,000 individual members and 75 affiliated Member Associations representing water quality professionals around the world. Since 1928, WEF and its members have protected public health and the environment. As a global water sector leader, our mission is to connect water professionals; enrich the expertise of water professionals; increase the awareness of the impact and value of water; and provide a platform for water sector innovation. To learn more, visit www.wef.org.

Prepared by the **Peracetic Acid Disinfection—Implementation Considerations for Water Resource Recovery Facilities** Task Force of the **Water Environment Federation**

Kati Bell, Ph.D., P.E., BCEE, *Chair*
Philip Block, Ph.D., *Vice-Chair*

Manuela Antonelli, Ph.D.
Robert Bastian
Somnath Basu
Keith Bourgeous
Bryan Buchanan
Arturo Burbano
Andrew Burnham
Joanne Carpenter
Daniela Castañeda
Cody Charnas, P.E.
Vince Chiu, P.E., M.E.
Shay Coombs
Ivan Cooper
Allegra da Silva, Ph.D., P.E.
Laura Dominguez Henao, Ph.D.
Bertrand Dussert
Gregory Eyerly
Richard Finger
Nicola Fontaine
Phyllis Frank
Bob Freeborn
Denise Funk, P.E., BCEE
George Garden, P.E., BCEE
Alberto Garibi
Ronnie Gehr
Josh Goldman
Theresa Gregonia
Ashley Griffin, P.E.
Amy Borello Gruss, Ph.D.
Philip Harvey
Greg Heath
Carl Hennagir
Brian Hilts, P.E.
Ronald Hofmann
Melanie Holmer
Gary Hunter
Joseph Husband
Joseph Jacangelo, Ph.D., REHS
Matthew Jalbert
Chris Jepson
Samuel Jeyanayagam
Leonard Levine
Samuel Lowen
Joanne MacDougall
Eddie Machek
Roberta Maffettone, Ph.D.
Kyriakos Manoli, Ph.D.
Nicole McLellan, M.A.Sc.
Ian Myers
Ana Peña-Tijerina, Ph.D., P.E. BCEE
Reynold Dale Richwine, P.E.
Amy Rushton, P.E.
Hodon Ryu
Andrew Salveson
Domenico Santoro
Siva Sarathy, Ph.D., P.E.
Kenneth Schnaars, P.E.
Edyta Stec-Uddin
Nicole Stephens, P.E.
Sarah Stewart, P.E., PMP
Andrea Turolla, Ph.D.
Michael Watts, Ph.D., P.E.
Elisabeth Werth
Jacquelyn Wilson
Joseph Viciere
Yue Xu

Under the Direction of the **Disinfection and Public Health Subcommittee** of the **Technical Practice Committee**

2020

Water Environment Federation
601 Wythe Street
Alexandria, VA 22314-1994 USA
http://www.wef.org

Special Publications of the Water Environment Federation

The WEF Technical Practice Committee (formerly the Committee on Sewage and Industrial Wastes Practice of the Federation of Sewage and Industrial Wastes Associations) was created by the Federation Board of Control on October 11, 1941. The primary function of the Committee is to originate and produce, through appropriate subcommittees, special publications dealing with technical aspects of the broad interests of the Federation. These publications are intended to provide background information through a review of technical practices and detailed procedures that research and experience have shown to be functional and practical.

Contents

List of Figures

List of Tables

Preface

Many utilities in the United States and Canada are looking to peracetic acid (PAA) disinfection, a technology that has been used in Europe for more than a decade, as a part of a risk management decision to reduce or eliminate gas chlorine and other disinfection byproducts from their treatment processes. However, there is a paucity of information available about the evaluation and implementation of this technology. This special publication provides an overview of the chemistry and kinetics of PAA disinfection, along with information for design, permitting, and implementation of the technology.

This first edition of this manual was produced under the direction of Kati Bell.

Authors' and reviewers' efforts were supported by the following organizations:

Black & Veatch
Brown and Caldwell
Carollo
CDM Smith
CHEMetrics, Inc.
Enviro Tech
Fort Worth Water Department
Garver
Gulf Coast Waste Disposal Authority
Metro Vancouver
Metro Wastewater Reclamation District
PeroxyChem
Politecnico di Milano, Department of Civil and Environmental Engineering, Environmental Division Stantec
State of Tennessee, Department of Environment & Conservation, Division of Water Resources
Trinity River Authority
Trojan Technologies
University of Western Ontario

1

Introduction

Kati Bell, Robert Bastian, & Daniela Castañeda

1.0 PURPOSE

The purpose of this special publication is to address considerations for peracetic acid (PAA; CH_3CO_3H) disinfection systems designed to meet bacterial compliance for discharges from wastewater resource recovery facilities (WRRFs) to receiving waters. This manual does not address PAA disinfection for drinking water or for reuse applications that require more stringent bacterial inactivation standards (because there is more direct human contact with the treated effluents). This manual also does not address high-rate disinfection for treatment of wet weather flows that have not received full secondary biological treatment (e.g., primary effluents or blended flows). Currently, there is no other guidance available on the design, implementation, and operation of PAA disinfection for wet weather flows that have not received full secondary biological treatment, although the topic is briefly addressed in *Wet Weather Design and Operation in Water Resource Recovery Facilities* (Water Environment Federation, 2014).

This manual provides information and guidance on PAA disinfection for wastewater discharges that are regulated by criteria established to protect human health and the environment. When disinfection standards are met at their sources, public safety and water quality are protected. It is also important to consider factors that influence and potentially bias treatment efficacy; these factors include influent water quality, facility design, equipment reliability, and operator training. Thus, it is necessary to provide guidance on PAA disinfection of wastewater, which is a relatively new technology for the wastewater disinfection market. This manual is designed to be useful to a broad audience, including WRRF operators, engineers and designers, regulators, and the scientific community.

1.1 Perspective

Population growth and other increases on water demands for supply and recreational uses significantly increase the opportunity for human exposure to wastewaters being discharged into the environment. The effectiveness of natural safeguards, such as dilution and distance or time before contact or use, have been reduced as a result of wastewater effluents being discharged in more locations and at larger volumes. Domestic wastewaters carry human pathogens excreted in fecal discharges of infected individuals, and even treated effluents can affect sources of domestic water supply, recreational waters, and shellfish growing areas.

Many WRRFs have historically discharged effluent to streams that are designated as recreational waters, are tributaries of larger waterbodies (e.g., streams or lakes), or are used as water supply sources by downstream communities. Potable water supplies are often extracted from these waterbodies, physically and chemically treated, and distributed to customers. Although drinking water facilities provide additional disinfection before distribution, the only protection recreational users receive is through adequate disinfection of wastewater effluents, which is necessary to reduce potential transmission of infectious diseases when human contact is possible.

Disinfection, in the context of wastewater treatment as described in this manual, aims to reduce pathogen concentrations such that human health risks are acceptable. The objective of wastewater disinfection is a reduction in concentrations of viable, pathogenic microorganisms responsible for illnesses. To reduce risks associated with fecal contamination, several disinfection methods are available. Municipal WRRFs can use these methods to reduce the dissemination of pathogens to the environment and break the potential cycles of pathogen-associated infections.

In the 20th century, chlorination became standard for disinfecting treated wastewater effluents and has been key to public health successes.

With awareness of the environmental effects associated with disinfection practices in the 1960s and 1970s, regulators identified a need to balance the destruction of pathogenic microorganisms against the negative effects of disinfected effluents on biota in receiving water and the serious public health consequences of disinfection byproducts (Whitby & Scheible, 2004).

The deleterious effects of halogens on the environment, along with the long-term effects of halogenated disinfection byproducts (DBPs), have been well documented. In the 1970s, this prompted governments throughout North America (Environment Canada, 1978; U.S. EPA, 1976) to introduce rules on residual chlorine and DBPs that have resulted in the reduced use of chlorine as a disinfectant for wastewater (Riordan, 1979). The implementation of alternative disinfection methods, such as ultraviolet (UV) or ozone, was encouraged (Whitby & Scheible, 2004); however, the high capital costs associated with treatment methods, such as UV and ozone, is a limitation to many WRRFs. Thus, the low capital cost of retrofitting chlorine disinfection facilities has been a key driver in recent applications of PAA disinfection facilities in North America.

Although PAA has been used for nearly two decades in Europe, it only recently has been applied for municipal wastewater disinfection in North America. The first commercial PAA product approved for use in the United States by the U.S. Environmental Protection Agency (U.S. EPA) was in 2006, with the first application in 2012. Table 1.1 summarizes common drivers for PAA, which has generated greater interest in use as a disinfectant. Section 2.2 discusses the regulatory drivers in more detail.

1.2 History of Peracetic Acid Disinfection

Peracetic acid is an important industrial peracid because of its high oxidation potential (Klenk et al., 2005). The chemical was prepared for the first time during 1900 to 1910, and, soon after, its sterilizing and disinfecting properties were reported (Swern, 1949). However, it was not until the 1970s when interest in water and wastewater treatment applications emerged (Baldry, 1983; Baldry & French, 1989; Meyer, 1976). Currently, the largest user of PAA is the food industry. Peracetic acid is used as a disinfectant and directly applied to meat to reduce *Escherichia coli* (*E. coli*), *Salmonella*, and *Listeria*. It is also used as a disinfectant in fresh produce washing water, clean-in-place processes, on food processing equipment, and in pasteurizers (Banach et al., 2015; Orth, 1998).

In healthcare, PAA is used in the sterilization of endoscopes (Chenjiao et al., 2016) and for the disinfection of surfaces in hospitals (Boyce, 2016). In industrial production of fine chemicals, PAA is used for epoxidation,

TABLE 1.1 Drivers for PAA disinfection in wastewater treatment.

Driver	Description
Capital cost	High capital costs are associated with alternative disinfection technologies such as UV or ozone disinfection; PAA can typically be implemented in existing structures provided adequate contact time is available.
Ease of operations and maintenance	Generally, no quenching is necessary, eliminating the need for a secondary chemical (analogous to dechlorination); it is notable that there are sometimes site-specific technical reasons for quenching. Longer shelf life decreases the frequency of chemical deliveries. Increased worker safety compared to chlorine gas.
Disinfection efficacy	PAA is a strong oxidizer and may eliminate some constituents of concern, such as endocrine-disrupting compounds. Ability to meet disinfection compliance under high total suspended solids or other challenging treatment conditions.
Regulatory requirements	Beneficial for increasingly stringent DBP limits because it does not produce halogenated DBPs. No risk management plan is required.
Toxicity	Low aquatic toxicity and fast decomposition kinetics.
Flexibility to retrofit	Ease of repurposing existing infrastructure, such as chlorine contact structures.

hydroxylation, and for the conversion of ketones to esters, for instance (Kaur & Kishore, 2014). Total chlorine-free processing of pulp uses distilled PAA at approximately 40% concentration as a bleaching agent when high brightness is required (Germgård, 2009). PAA is also used as a biocide in paper mill process waters (Jakara et al., 2000; Kiuru, 2011). With respect to wastewater disinfection, enough research had been conducted for a review over a decade ago, in which the potential advantages and disadvantages of PAA for wastewater disinfection was outlined (Kitis, 2004). A more recent review was published that has updated and extended water-treatment-related information regarding PAA (Luukkonen & Pehkonen, 2017).

1.2.1 Indicator Bacteria

The rationale for using indicator organisms as the basis for microbiological criteria is that it is not feasible, nor practical, to detect and enumerate every pathogen in water; therefore, it is not possible to assess and understand the specific risk to public and environmental health posed by each pathogen or its quantity per unit volume at any given time. This is because risk depends on the infectivity and invasiveness of a pathogen and the innate and acquired immunity of individuals contacting the water. In addition, only certain waterborne pathogens can be detected reliably and easily in water, and some cannot be detected at all (WHO, 1996). Thus, the best indicators of fecal contamination will be those that are universally present in large numbers in the feces of humans and warm-blooded animals, are readily detected by simple methods, do not grow in natural waters, and persist in water and can be removed by wastewater treatment at rates similar to waterborne pathogens.

In 1968, the National Technical Advisory Committee (NTAC) translated the previously established total coliform level of 2300/100 mL (Stevenson, 1953) to 400 fecal coliforms/100 mL based on a ratio of total to fecal coliform, and then halved that number to 200 fecal coliforms/100 mL (U.S. EPA, 2012). U.S. EPA recommended the NTAC criteria for recreational waters again in 1976, even though the criteria had been criticized for several issues related to the design of the U.S. Public Health Service (USPHS) studies of public health risks associated with bacteria concentrations in recreational water, and the limited amount of epidemiological data and data quality. The 1976 U.S. EPA criterion for bacteria in primary recreational waters required that fecal coliform not exceed a geometric mean of 200 organisms/100 mL, and that no more than 10% of the total number of samples, taken during any 30-day period, exceed 400 fecal coliforms/100 mL (U.S. EPA, 1976). By 1986, as more data became available, U.S. EPA recommended *E. coli* and enterococci be used for assessing microbiological water quality in recreational waters because concentrations of these organisms are more strongly correlated with swimming-associated gastroenteritis rates (U.S. EPA, 1986). It should be noted that many states only apply criteria seasonally to protect human health when human contact would occur.

In the United States, some states questioned whether they should adopt the 1986 recommendations for *E. coli* or enterococci for setting water quality standards, and whether it was necessary to change their programs if the estimation of disease risk to swimmers had not significantly improved. Because of new studies and data, U.S. EPA took the position that *E. coli* and enterococci were better indicators of public health risk in recreational waters than fecal coliforms; results from epidemiological evidence firmly linked *E. coli* and enterococci levels to swimming-related illness (Cabelli, 1983; Dufour, 1984).

When developing criteria based on *E. coli* and enterococci, U.S. EPA proposed criteria that were no more stringent than the 200 fecal coliforms/100 mL first recommended in 1968. Instead, they represented the disease risk estimated for swimmers at freshwater and marine beaches where swimmers would be exposed to the maximum fecal coliform concentrations. The 1986 criteria were calculated to represent the ambient condition of the waterbody necessary to protect the designated use of primary contact recreation. These values were selected to carry forward the same level of illness risk associated with U.S. EPA's previous criteria recommendations to protect the primary contact recreation use, which were for fecal coliforms (U.S. EPA, 1976).

The 1986 criteria also carried a single sample maximum (SSM) component, which was computed using the geometric mean values and corresponding observed variances in the fecal indicator bacteria obtained from water quality measurements taken during the epidemiological studies from the late 1970s and early 1980s. Four different SSM values (to be used with different recreational use intensities) were provided and correspond to different percentiles of the water quality distribution around the geometric mean. The 1986 criteria for marine and fresh water were based on different water quality values and associated illness rates because the marine and freshwater epidemiological studies reported different geometric mean values for indicator bacteria associated with the water quality corresponding to U.S. EPA's fecal coliform criteria recommendation.

Epidemiological studies have been used for decades to evaluate how fecal indicator bacteria concentrations are associated with health effects of primary contact recreation on a quantitative basis. The 1986 criteria are supported by studies conducted by U.S. EPA in the late 1970s and early 1980; in these studies, enterococci and *E. coli* exhibited the strongest correlations to swimming-associated gastroenteritis. Both indicators are used in epidemiological studies throughout the world, including in the European Union and Canada (Council of the European Communities, 2006). The World Health Organization (WHO) recommends the use of enterococci as water quality indicators for recreational waters (WHO, 2003). Meta analyses and systematic reviews of epidemiological studies conducted worldwide indicate that these indicators generally provided substantial improvements over those favored previously, such as total and fecal coliforms (Prüss, 1998; Wade et al., 2003; Zmirou, Pena, Ledrans, & Leterte, 2003). Thus, when U.S. EPA most recently updated its recreational water quality criteria (RWQC) in 2012, enterococci and *E. coli* were again recommended for fresh water, and enterococci as the indicator to be measured in both marine and fresh water. The most recent U.S. EPA (2012) RWQC offers two sets of recommended numeric concentration thresholds, either of which would be protective for primary contact in recreational waters.

1.2.2 Enumeration Methods for Indicator Bacteria

Indicators of fecal contamination are detected and enumerated using a variety of methods. Fecal indicator bacteria are used as the basis of monitoring in discharge permits. These indicators can be enumerated using various analytical methods, including those in which the organisms are grown (cultured) and those in which their DNA is extracted from an environmental sample, then amplified and quantified (using quantitative polymerase chain reaction [qPCR]). These different enumeration methods result in method-specific units and values. One culture-based method, membrane filtration, yields the number of colonies that arise from bacteria captured on the membrane filter per volume of water filtered; one colony can be produced from one or several cells (clumped cells in the environmental sample) and results are expressed in units such as colony-forming units per volume (e.g., CFU/100 mL). Other culture-based measurements, based on use of a defined substrate method, produce a most probable number (MPN) per volume. Bacterial density MPNs are based on a combination of positive and negative test tube results read from an MPN table (U.S. EPA, 1978). The IDEXX Quanti-Tray® is a commercial kit that provides MPN results for coliforms, *E. coli*, enterococci, *Pseudomonas aeruginosa*, and heterotrophic plate counts. Results from qPCR analyses are reported in units calculated based on target DNA sequences from test samples relative to those in calibrator samples that contain a known quantity of target organisms (Haugland, Siefring, Wymer, Brenner, & Durfour, 2005; Wade et al., 2010).

There are challenges to interpreting results from each of these enumeration techniques because results from each analytical technique can vary depending on a number of factors. For example, culture-based methods depend on the metabolic state (i.e., viability and activity) of target organisms for effective enumeration. In contrast, qPCR-based approaches detect specific sequences of DNA that have been extracted from a water sample, and results contain sequences from both viable and nonviable forms of the targeted indicator.

1.3 Water Quality Criteria as the Basis for Disinfection

Public health agencies worldwide have long understood the relationship between fecal contamination in surface waters and the associated human health risks. Because of the difficulties in identifying the specific origin of illnesses associated with fecal contamination, as early as the 1960s, USPHS recommended using fecal coliform bacteria as an indicator for human health risks associated with primary contact. This recommendation was based on studies that reported a detectable health effect when total coliform densities exceeded 2300 CFU/100 mL (Stevenson, 1953).

While the correlations between fecal coliform bacteria and waterborne illnesses have been documented, interestingly, it is also known that most strains of fecal indicator bacteria (i.e., those that are used for monitoring) are not pathogenic. It is the presence of pathogenic organisms such as viruses, pathogenic bacteria, and protozoa that cause these illnesses. Although many species of fecal coliform bacteria are not pathogenic, these microorganisms demonstrate characteristics as indicators of fecal contamination (i.e., often of fecal origin and simple methods of detection) and indirectly indicate the potential presence of fecal pathogens capable of causing illnesses. Scientists recognize that the use of an indicator is not a perfect method for detecting the presence of all of the numerous pathogens that cause illnesses associated with exposure to surface waters where wastewater is discharged. However, use of these indicators is supported by epidemiological studies on human health relationships, and this approach overcomes issues associated with pathogen-specific enumeration methods for environmental waters (U.S. EPA, 2012). Further, indicator organisms have and are used as a criterion forming the basis of regulatory frameworks for wastewater disinfection.

The 2012 U.S. EPA RWQC, presented in Table 1.2, protect the public from exposure to harmful levels of pathogens. Illness rates that U.S. EPA recommended are based on the National Epidemiological and Environmental Assessment of Recreational Water definition of "gastrointestinal illness", which is limited to illnesses that exhibit a fever. This study allowed U.S. EPA to provide better estimates of risk based on the new data. These recommendations have been issued as guidance to states, territories, and authorized tribes for use in developing water quality standards to protect swimmers from exposure to water that contains organisms that indicate the presence of fecal contamination. It should be noted the SSM has been dropped in the 2012 U.S. EPA RWQC in favor of a statistical threshold value (STV); new criteria are comprised of both a geometric mean and the STV. For a set of criteria values, U.S. EPA computed the STV based on water quality distribution observed during U.S. EPA's epidemiological studies. The STV approximates the 90th percentile of the water quality distribution and is intended to be a value that should not be exceeded by more than 10% of samples used to calculate the geometric mean. Because fecal indicator bacteria are highly variable in ambient waters, distributional estimates are more robust than single-point estimates (U.S. EPA, 2012).

U.S. EPA has initiated activities to advance the development of a coliphage (a viral indicator) to protect public health from water sources that have been influenced by fecal contamination or wastewater discharges

TABLE 1.2 Recommended recreational water quality criteria (U.S. EPA, 2012).

	Estimated illness rate: 36 per 1000 primary contact recreators		Estimated illness rate: 32 per 1000 primary contact recreators	
Criteria elements	**Magnitude**		**Magnitude**	
Indicator	Geometric mean (CFU/100 mL)	STV (CFU/100 mL)	Geometric mean (CFU/100 mL)	STV (CFU/100 mL)
Enterococci—marine and fresh	35	130	30	110
E. coli—fresh	126	410	100	320

Note. Duration and frequency: The waterbody geometric mean should not be greater than the selected geometric mean magnitude in any 30-day interval. There should not be greater than a 10% excursion frequency of the selected STV magnitude in the same 30-day interval.

as part of the 5-year review process of the 2012 RWQC. In April 2015, U.S. EPA published a literature review on the state of the science describing the usefulness of coliphage as an indicator (https://www.epa.gov/wqc/microbial-pathogenrecreational-water-quality-criteria#coliphage). Although the U.S. EPA literature review asserted that coliphages are generally an equally good indicator of fecal contamination as the agency's currently recommended criteria for *E. coli* and enterococci, U.S. EPA has yet to issue water quality criteria.

U.S. EPA held a Coliphage Experts Workshop in March 2016 as part of its ongoing efforts to build the scientific basis for developing coliphage-based water quality criteria. U.S. EPA convened a group of 12 internationally recognized experts on the state of the science of coliphage and its usefulness as a viral indicator for the protection of public health in recreational waters. Experts represented a spectrum of perspectives from academia, federal agencies (U.S. EPA, Centers for Disease Control and Prevention, U.S. Food and Drug Administration), and the wastewater industry. Agenda discussion topics included the need for a viral indicator, coliphage as a predictor of gastrointestinal illnesses, how coliphage may be useful as an indicator of wastewater treatment performance, male-specific versus somatic coliphage, a systematic literature review of viral densities, and future research. The accompanying Coliphage Experts Workshop Proceedings document details topics and findings from the workshop. Additionally, the document provides expert written responses to charge questions provided before the workshop (U.S. EPA, 2016; https://www.epa.gov/wqc/microbial-pathogenrecreational-water-quality-criteria).

To further the discussion on coliphage criteria, U.S. EPA also recently validated and published two draft methods for measurement of coliphage: methods 1642 and 1643 (U.S. EPA, 2018). Draft U.S. EPA Method 1643 is a single agar layer procedure to detect and enumerate male-specific and somatic coliphages. This culture-based method can be used for monitoring secondary (no disinfection) wastewater matrices under the Clean Water Act (CWA). The validation study reflects the results of a multi-laboratory validation study of U.S. EPA Method 1602 for 100-mL secondary (no disinfection) wastewater samples; however, use of this method for disinfected samples has not been validated. To establish concentrations of coliphage in disinfected wastewater and in recreational waters, draft U.S. EPA Method 1642 can be used to concentrate, detect, and enumerate male-specific and somatic coliphage. This method has been modified from U.S. EPA Method 1602 to include dead-end hollow-fiber ultrafiltration to concentrate large sample volumes (>2 L) required for recreational water monitoring to a final sample volume of approximately 200 mL, which can then be assayed for both male-specific and somatic coliphages using the single agar layer procedure (draft U.S. EPA Method 1643).

In 2017, U.S. EPA published its 5-year review of its 2012 RWQC, as required by the Beaches Environmental Assessment and Coastal Health (BEACH) Act amendments to CWA section 304(a)(9)(B). In conducting the review, U.S. EPA considered several factors, including availability and evaluation of new science, review of information related to the underlying science used to develop the 2012 RWQC, additional implementation support needs, and perceived barriers to state adoption. Based on review of existing criteria and scientific developments since publication of the 2012 RWQC, U.S. EPA decided not to revise the 2012 RWQC. U.S. EPA stated that further research and analyses are needed and that it has been working with the environmental public health community as the agency moves forward with its ongoing research, including a recent literature review of coliphage densities in raw wastewater and ambient waters (Eftim et al., 2017; Nappier et al., 2019), which will be used to inform the derivation of a coliphage RWQC.

2.0 REGULATORY CONSIDERATIONS

Globally, the driver for disinfection of wastewater effluents is the protection of human health from recreational contact with potentially pathogenic organisms in waterbodies and/or to protect water supplies from being contaminated with pathogenic microorganisms. In the United States and Canada, wastewater discharges are regulated through a framework that is implemented at a state or province level. Although the regulatory approach varies globally, in general there are criteria, such as those set by WHO (2003), that provide guidance on minimum treatment requirements for wastewater that are generally consistent with the U.S. EPA criteria described in Section 1.2.1.

2.1 Regulatory Frameworks in the United States and Canada

In the United States, CWA set a national goal "to restore and maintain the chemical, physical, and biological integrity of the Nation's waters", with an interim goal that all waters be "fishable and swimmable" where possible. The CWA specifies that all point-source discharges into the nation's waters are unlawful unless authorized by a National Pollution Discharge Elimination System (NPDES) permit and sets baseline, technology-based controls for municipalities and industries. It requires all dischargers to meet additional stricter pollutant controls where needed to meet water quality targets and requires federal approval of these standards. The CWA also provides the authority for U.S. EPA to establish specific criteria for protecting the public from exposure to harmful levels of pathogens while participating in water-contact activities such as swimming, wading, and surfing in waters designated for such uses. Thus, it is recreational water quality uses that drive permit limits for disinfection.

In Canada, the Canadian Environmental Protection Act, 1999 (CEPA 1999) is the primary element of the legislative framework for protecting the Canadian environment and human health (Environment Canada, 1999). A key aspect of CEPA 1999 is the prevention and management of risks posed by toxic and other harmful substances, including microorganisms. The minister of the environment is accountable to the Canadian Parliament for the administration of all of CEPA 1999. The minister of the environment and the minister of health jointly administer the task of assessing and managing the risks associated with toxic substances. Efforts taken under CEPA 1999 are complemented by actions taken under other federal acts administered by the minister of the environment. The Fisheries Act, administered by the minister of the environment on behalf of the minister of fisheries and oceans, includes provisions to prevent pollution of waters inhabited by fish. Water resources and their environmental quality are also managed through the Canada Water Act.

2.2 Regulatory Drivers for Peracetic Acid

In 2015, U.S. EPA issued updated national recommended water quality criteria for human health for 94 chemical pollutants to reflect the latest scientific information and U.S. EPA policies. The update included new fish consumption rates and limits for some disinfection byproducts that are more restrictive than the previously published criteria. These water quality criteria serve as guidance for states and tribes authorized to establish water quality standards under the CWA. In addition to disinfection byproducts, individual states are faced with increasingly stringent nutrient limits that have a direct effect on chlorine-based disinfection. Similar activities require WRRFs in Canada to review their current practices of chlorine disinfection.

In 2006, Environment Canada began work on a national strategy, under direction of the Canadian Council of Ministers of the Environment (CCME), to provide additional requirements for management of wastewater effluents. In 2007, CCME released a "Draft Canada-wide Strategy for the Management of Municipal Wastewater Effluent". At the same time, Environment Canada published a "Proposed Regulatory Framework for Wastewater" to explain how the Canada-wide strategy would be implemented. A draft regulation based on the CCME strategy was released for comment in 2010 and, on July 28, 2012, the Wastewater Systems Effluent Regulations (WSER) were published in the *Canada Gazette* (Environment Canada, 2012). The regulations fall under the Fisheries Act, which prohibits unauthorized deterioration, disruption, and destruction of fish habitat.

The WSER represent the first national regulations in Canada that specifically address municipal WRRF effluents that are regulated under provincial authorities. In the Canadian WSER, these parameters include un-ionized ammonia, acute lethality, and total residual chlorine (TRC). As a result, facilities that have historically used chlorination may be faced with variable effluent ammonia and nitrites when solids retention time is increased to improve treatment for ammonia. *White's Handbook of Chlorination and Alternative Disinfectants* (Black & Veatch, 2010) contains a detailed discussion of the effects of these ions on chlorination and the disinfection process control using chlorine. Thus, with respect to WRRFs, emerging criteria and new regulations may impose limits for parameters in final effluents that affect disinfection process control for chlorination. These challenges, coupled with the need to add dechlorination to meet toxicity or TRC limits, are drivers for many facilities to investigate PAA disinfection as a solution to the disinfection challenges associated with chlorination.

3.0 ORGANIZATION OF THE PUBLICATION

The chapters of this publication are organized such that the reader can develop an understanding of the theory of PAA disinfection before reading about system design, storage and feed equipment requirements, construction, operations, and compliance considerations. With this background, the reader will have a better grasp of the content of the chapters that provide details on engineering design, construction, and operations of these systems.

Chapter 2 provides information on PAA chemical properties and manufacturing along with considerations for health and safety. Chapter 3 presents the mechanisms and kinetics of disinfection. Chapter 4 includes a discussion on the methods of PAA analysis and relevant process monitoring tools for wastewater disinfection applications. Chapter 5 describes considerations for effluent toxicity and calculation of allowable residuals in disinfected effluents. Chapter 6 includes a description of the disinfection byproducts that are associated with PAA and a discussion around the potential that PAA may also reduce the activity and concentrations of some contaminants of emerging concern. Chapter 7 provides an overview of PAA performance testing, from bench-scale to full-scale. Chapter 8 describes the regulatory considerations surrounding PAA permitting and implementation, along with a summary of information from states that have allowed or permitted use of PAA.

Chapter 9 provides an overview of design and process control considerations, and Chapter 10 covers the construction, installation, and commissioning of PAA disinfection systems. Chapter 11 summarizes the operational considerations for wastewater PAA disinfection systems along with strategies for meeting continuous compliance. Chapter 12 discusses the economics of PAA disinfection, considering that there are non-economic benefits of PAA that are not captured within the capital and operating costs. Chapter 13 provides case study examples that demonstrate information presented in previous chapters. Although it is not the primary intent of this publication to address other applications of PAA in wastewater treatment (e.g., reuse, drinking water, and biosolids treatment), the fundamental information presented in this manual and the related case studies do provide relevant background for those applications. Because of the interest in these PAA applications, Chapter 14 provides a brief overview of other PAA applications related to wastewater treatment.

4.0 REFERENCES

Baldry, M. G. C. (1983). The bactericidal, fungicidal and sporicidal properties of hydrogen peroxide and peracetic acid. *The Journal of Applied Bacteriology, 54*(3), 417–423.

Baldry, M. G. C., & French, M. S. (1989). Disinfection of sewage effluent with peracetic acid. *Water Science & Technology, 21*(3), 203–206.

Banach, J. L., Sampers, I., Van Haute, S., & van der Fels-Klerx, H. (2015). Effect of disinfectants on preventing the cross-contamination of pathogens in fresh produce washing water. *International Journal of Environmental Research and Public Health*, *12*(8), 8658–8677.

Black & Veatch. (2010). *White's handbook of chlorination and alternative disinfectants* (5th ed.). (In Dominic M. Desiderio & Nico M. M. Nibbering, Eds.). Hoboken, NJ: Wiley & Sons.

Boyce, J. M. (2016). Modern technologies for improving cleaning and disinfection of environmental surfaces in hospitals. *Antimicrobial Resistance and Infection Control, 5*(1), 1–10.

Cabelli, V. J. (1983). *Health effects criteria for marine recreational waters* (EPA-600/1-80-031). Washington, DC: U.S. Environmental Protection Agency.

Chenjiao, W., Hongyan, Z., Qing, G., Xiaoqi, Z., Liying, G., & Ying, F. (2016). In-use evaluation of peracetic acid for high-level disinfection of endoscopes. *Gastroenterology Nursing*, *39*(2), 116–120.

Council of the European Communities. (2006). Directive 2006/7/EC of the European Parliament and of the Council of 15th February 2006 concerning the management of bathing water quality and repealing Directive 76/160/EEC. *Official Journal of the European Communities,* L64, 37–51.

Dufour, A. P. (1984). *Health effects criteria for fresh recreational waters* (EPA-600/1-84-004). Washington, DC: U.S. Environmental Protection Agency.

Eftim, S. E., Hong, T., Soller, J., Boehm, A., Warren, I., Ichida, A, & Nappier, S. P. (2017). Occurrence of norovirus in raw sewage: A systematic literature review and meta-analysis. *Water Research*, 111, 366–374.

Environment Canada. (1978). *Wastewater disinfection in Canada* (EPS 3-WP-78-4). Gatineu, Quebec, Canada: Environmental Protection Service, Environment Canada.

Environment Canada. (1999). Canadian Environmental Protection Act, 1999. Retrieved from http://www.ec.gc.ca/lcpe-cepa/default.asp?lang=En&n=CC0DE5E2-1&toc=hide (accessed Jan 2015).

Environment Canada. (2012). Wastewater Systems Effluent Regulations. Retrieved from http://laws-lois.justice.gc.ca/eng/regulations/SOR-2012-139/FullText.html (accessed Jan 2015).

Germgård, U. (2009). Production of bleaching chemicals at the mill. In M. Ek, G. Gellerstedt, & G. Henriksson (Eds.), *Pulping chemistry and technology*, Vol. 2 (Chapter 11, pp. 277–297). Berlin, Germany: Walter de Gruyter.

Haugland, R. A., Siefring, S. C., Wymer, L. J., Brenner, K. P., & Durfour, A. (2005). Comparison of Enterococcus measurements in fresh water at two recreational beaches by quantitative polymerase chain reaction and membrane filter culture analysis. *Water Research, 39*(4), 559–568.

Jakara, J., Paren, A., Hukkanen, P., & Autio, P. (2000, April 3–6). Effect of peracetic acid in fine paper production. 54th APPITA Annual General Conference, Melbourne, Australia.

Kaur, N., & Kishore, D. (2014). Peroxy acids: Role in organic synthesis. *Synthetic Communications,* 44, 721–747.

Kitis, M. (2004). Disinfection of wastewater with peracetic acid: A review. *Environment International,* 30, 47–55.

Klenk, H., Götz, P. H., Siegmeier, R., & Mayr, W. (2005). *Organic peroxy compounds. Ullmann's encyclopedia of industrial chemistry.* New York, NY: Wiley & Sons.

Kiuru, J. (2011). *Interactions of chemical variations and biocide performance at paper machines* (p. 46) (Ph.D. Thesis). Aalto University, Helsinki, Finland.

Luukkonen, T., & Pehkonen, S. O. (2017). Peracids in water treatment: A critical review. *Critical Reviews in Environmental Science and Technology, 47*(1), 1–39.

Meyer, E. (1976). Disinfection of sewage waters from rendering plants by means of peracetic acid. *Journal of Hygiene Epidemiology, Microbiology, and Immunology, 21*(3), 266–273.

Nappier, S. P., Hong, T., Ichida, A., Goldstone, A., & Eftim, S. (2019). Occurrence of coliphage in raw wastewater and in ambient water: A meta-analysis. *Water Research*, 153, 263–273.

Orth, R. (1998). The importance of disinfection for the hygiene in the dairy and beverage production. *International Biodeterioration & Biodegradation,* 41, 201–208.

Prüss, A. (1998). Review of epidemiological studies on health effects from exposure to recreational water. *International Journal of Epidemiology, 27*(1), 1–9.

Riordan, C. (1979). Perspectives on wastewater disinfection—A view from headquarters. In *Progress in Wastewater Disinfection Technology* (EPA-600/9-79-018). Cincinnati, OH: U.S. Environmental Protection Agency.

Stevenson, A. H. (1953). Studies of bathing water quality and health. *American Journal of Public Health and the Nation's Health,* 43, 429–538.

Swern, D. (1949). Organic peracids. *Chemical Reviews*, 45, 1–68.

U.S. Environmental Protection Agency. (1976). *Disinfection of wastewater—Task force report* (EPA-430/9-75-013). Washington, DC: U.S. Environmental Protection Agency.

U.S. Environmental Protection Agency. (1978). *Microbiological methods for monitoring the environment: Water and wastes* (EPA-600/8-78-017). Cincinnati, OH: U.S. Environmental Protection Agency, Office of Research and Development.

U.S. Environmental Protection Agency. (1986). Ambient Water Quality for Bacteria—1986 (440/5-84-002). Washington, DC: U.S. Environmental Protection Agency.

U.S. Environmental Protection Agency. (2012). *Recreational water quality criteria* (EPA-820/F-12-058). Washington, DC: U.S. Environmental Protection Agency.

U.S. Environmental Protection Agency. (2016). 2016 Coliphage Experts Workshop. https://www.epa.gov/sites/production/files/2019-03/documents/coliphage-experts-workshop-factsheet-2016.pdf

U.S. Environmental Protection Agency. (2018). *Method 1643: Male-specific (F+) and somatic coliphage in secondary (no disinfection) wastewater by the single agar layer (SAL) procedure* (EPA-820/R-18-003). Washington, DC: U.S. Environmental Protection Agency, Office of Water.

Wade, T. J., Pai, N., Eisenberg, J. N. S., & Colford, J. M., Jr. (2003). Do U.S. Environmental Protection Agency water quality guidelines for recreational waters prevent gastrointestinal illness? A systematic review and meta-analysis. *Environmental Health Perspectives, 111*(8), 1102–1109.

Wade, T. J., Sams, E., Brenner, K. P., Haugland, R., Chern, E., Beach, M., … Dufour, A. P. (2010). Rapidly measured indicators of recreational water quality and swimming-associated illness at marine beaches: A prospective cohort study. *Environmental Health, 9*(66).

Water Environment Federation. (2014). *Wet weather design and operation in water resource recovery facilities*. Alexandria, VA: Water Environment Federation.

Whitby, G. E., & Scheible, O. K. (2004). The History of UV and wastewater. *IUVA News, 6*(3), 15–26.

World Health Organization. (1996). *Guidelines for drinking-water quality. Volume 2: Health criteria and other supporting information* (2nd ed.). Geneva, Switzerland: World Health Organization.

World Health Organization. (2003). *Guidelines for safe recreational water environments: Volume 1 Coastal and fresh waters*. Retrieved from http://www.who.int/water_sanitation_health/bathing/srwe1/en/ (accessed Jan 2015).

Zmirou, N., Pena, L., Ledrans, M., & Leterte, A. (2003). Risks associated with the microbiological quality of bodies of fresh and marine water used for recreational purposes: Summary estimates based on published epidemiological studies. *Archives of Environmental Health, 58*(11), 703–711.

2

Fundamentals of Peracetic Acid Chemistry and Manufacturing

Melanie Holmer, Andrea Turolla, Kati Bell, Manuela Antonelli, Daniela Castañeda, Alberto Garibi, & Jacquelyn Wilson

1.0 INTRODUCTION

As described in Chapter 1, peracetic acid (PAA), also known as *peroxyacetic acid*, is a chemical gaining a great deal of interest in North America because of its ability to provide bacterial inactivation performance at costs competitive to other mature disinfection technologies. Peracetic acid is a corrosive, clear, colorless liquid with a pungent (vinegar) odor. It is also a very strong oxidizer and possesses many of the characteristics of hydrogen peroxide (H_2O_2). Peracetic acid has a pH of less than 2. Many metals, polymers, and elastomers will experience degradation when exposed to PAA. These properties are important considerations in the selection of materials for design, construction, installation, and operation of PAA chemical facilities.

1.1 Physical and Chemical Properties of Peracetic Acid

Peracetic acid belongs to the group of percarboxylic acids (also referred to as *organic peracids*), namely chemical compounds whose general molecular structure is R-COOOH. Within this group, PAA is a short-chain compound, for which the functional group R is a methyl group (CH_3), hence its chemical formula is CH_3COOOH (Block, 2001). Peracetic acid solutions exist in equilibrium concentrations with water, hydrogen peroxide, and acetic acid, as shown in Figure 2.1.

The oxidation potential of PAA is higher than that of other chemicals used in water and wastewater treatment—being nearly as high as that of ozone (Figure 2.2)—which accounts for its ability to both disinfect and oxidize organic chemicals. Because of its reactivity, PAA does not persist in the environment and breaks down into acetic acid (vinegar) and hydrogen peroxide, which subsequently decompose to oxygen and water. Peracetic acid and hydrogen peroxide have low octanol-water partition coefficients (K_{OW}) (0.3 and 0.4, respectively) and low sediment adsorption coefficients, so bioaccumulation in aquatic organisms or in sediments is highly unlikely (European Centre for Ecotoxicology and Toxicology of Chemicals, 2001).

In general, PAA equilibrium solutions are stable and can maintain their strength for months if stored in adequate conditions. However, PAA naturally decomposes over time, with the process accelerated in the presence of transition metal ions (such as iron) or when its temperature increases beyond its self-accelerated decomposition temperature (SADT). This occurs by the release of oxygen from PAA, decomposing into acetic acid (CH_3COOH), via two exothermic reactions: the first being spontaneous or catalyzed by metals

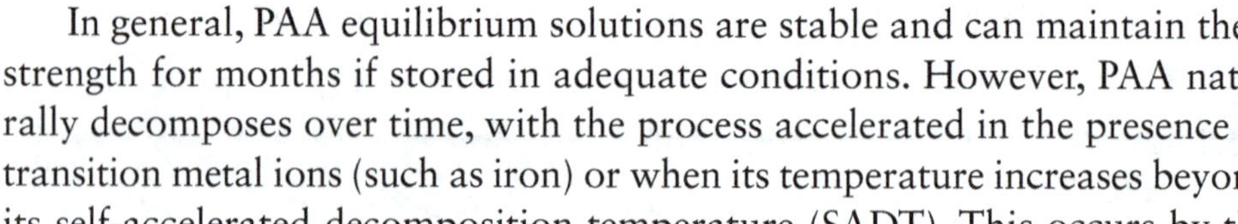

FIGURE 2.1 Equilibrium solution of commercially available solution-grade peracetic acid (graphic courtesy of PeroxyChem).

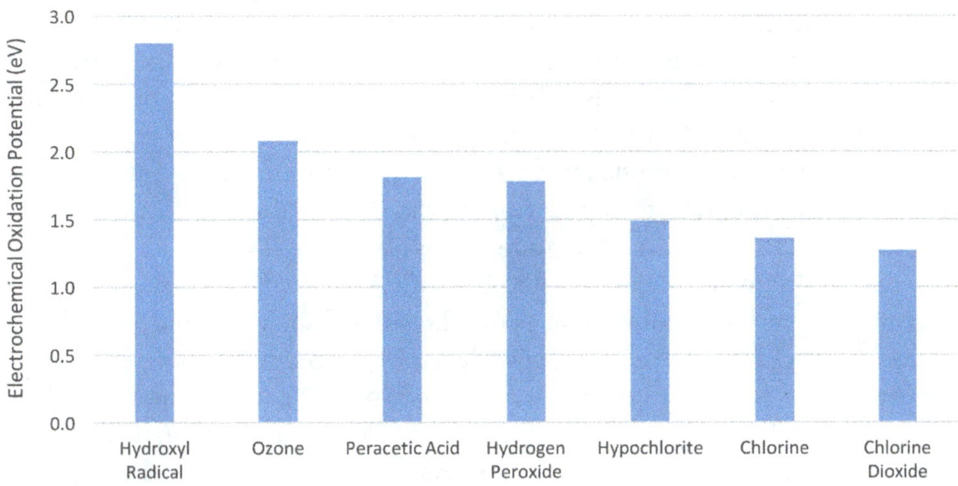

FIGURE 2.2 Redox potential of commonly used water and wastewater treatment chemicals.

and the second being hydrolytic (Yuan, 1997; Yuan, Ni, & Van Heiningen, 1997; Zhao et al., 2007; Zhao, Cheng, Hao, & Liu, 2008), as follows:

$$2CH_3COOOH(+Metal) \rightarrow 2CH_3COOH + O_2(+Metal) \quad (2.1)$$

$$CH_3COOOH + H_2O \rightarrow CH_3COOH + H_2O_2 \quad (2.2)$$

In addition to the reaction reported in Equation 2.1, another minor thermic decomposition reaction resulting in the generation of gaseous products was proposed in the past (Drimus & Matasa, 1966):

$$CH_3COOOH \rightarrow CH_3COH + CO_2 \quad (2.3)$$

The influence of temperature and pH on the decomposition of PAA was assessed in several previously published works under various conditions of PAA concentration and composition of the aqueous matrix. Literature unambiguously indicates that the kinetics of decomposition accelerate at increasing temperature (*inter alia*, Pedersen, Meinelt, & Straus, 2013; Kunigk et al., 2001; Kunigk, Silva, & Jurkiewicz, 2012). Concerning the effect of pH on the decomposition of PAA, faster kinetics were generally observed at higher pH values. Furthermore, the prevalence of the decomposition mechanisms, spontaneous or hydrolytic, depends on the pH (European Centre for Ecotoxicology and Toxicology of Chemicals, 2001; Yuan, 1997; Yuan et al., 1997; Zhao, Cheng, Hao, & Liu, 2008). The decomposition is mostly driven by spontaneous reactions (Equation 2.1) in a pH range between 5.5 and 8.2. Within this range, the decomposition kinetics can be effectively described by a second-order kinetic model with respect to the total PAA concentration. The influence of hydrolytic reactions (Equation 2.2) increases with respect to spontaneous reactions for pH values between 8.2 and 9.0. Hydrolytic reactions can be modelled as a first-order with respect to the total PAA concentration.

Two aspects that significantly affect the stability of PAA are its concentration in solution and its application mode. Concerning the former, generally, the more PAA is diluted in aqueous solution, the higher its decomposition rate. As an example, while 1 to 2% of PAA undergoes decomposition in about 1 month in a solution at 40%, the half-life of 1% solution is about 6 days (Block, 2001). A slight dependence of decomposition rate on PAA concentration was also reported at values typical of wastewater disinfection (i.e., 1–10 mg L^{-1}) (Domínguez Henao, Delli Compagni, Turolla, & Antonelli, 2018; Pedersen et al., 2013). As for the influence of the application mode on PAA stability, the commercial products are commonly dosed as manufactured without intermediate dilution steps when PAA solutions are used at the industrial scale, in the view of minimizing the overall loss of PAA by decomposition (Kitis, 2004).

The concentration of each constituent in solution depends on the application and the manufacturing process; 10 to 15% of PAA (w/w) is a common composition for commercial products intended for many industrial applications, including wastewater disinfection (European Centre for Ecotoxicology and Toxicology of Chemicals, 2001). Later in the chapter, Table 2.2 summarizes the composition of commercial PAA solutions registered with the U.S. Environmental Protection Agency (U.S. EPA) for wastewater disinfection in the United States. Main constituents examined include PAA, hydrogen peroxide, and acetic acid.

Table 2.1 lists the physical and chemical properties of PAA (European Centre for Ecotoxicology and Toxicology of Chemicals, 2001). These properties characterize the pure component of PAA, regardless of its concentration in solution. Because relevant variations occur among different available commercial products, further modifications are expected depending on the specific case. Therefore, consulting the appropriate safety data sheet (SDS) is strongly recommended.

2.0 PERACETIC ACID PRODUCTION

Although there are alternative methods for industrial production of the chemical, PAA used in wastewater disinfection is produced by a reaction between acetic acid and hydrogen peroxide. This is often achieved using a catalyst (Zhao, Zhang, Zhou, & Liu, 2008), according to the equation shown in Figure 2.1.

TABLE 2.1 Physical and chemical properties of component peracetic acid. Table adapted from European Centre for Ecotoxicology and Toxicology of Chemicals (2001).

Property	Value	Reference
Solubility in water at 20 °C (68 °F)	1000 g/kg	Swern (1970)
pK_a at 20 °C (68 °F)	8.2	Swern (1970)
Partition coefficient log K_{OW} (octanol/water) at 25 °C (77 °F)	0.3	European Centre for Ecotoxicology and Toxicology of Chemicals (2001)
Odor threshold	50 ppb	European Centre for Ecotoxicology and Toxicology of Chemicals (2001)
Henry's law constant at 25 °C (77 °F)	0.22 Pa·m^3/mol	O'Sullivan, Lee, Noone, & Heikes (1996)

2.1 Commercial Solution-Grade Peracetic Acid

The concentration of peracetic acid as the active ingredient can vary in specific formulations of PAA; the available PAA in solution is controlled by managing the initial concentration and amounts of hydrogen peroxide and acetic acid during the manufacturing process. Higher concentrations of PAA are prepared using higher strength hydrogen peroxide and varying the molar ratio of acetic acid to hydrogen peroxide. This is an important consideration in the selection of a PAA product for wastewater applications because each formulation introduces a different amount of acetic acid, which has an associated biological oxygen demand (BOD) for each milligram per liter (mg/L) of PAA dosed. Additional factors that impact PAA concentration include temperature, concentration and amount of sulfuric acid (catalyst), and reaction or equilibrium time (Zhao, Zhang, Zhou, & Liu, 2008).

2.1.1 Catalysts

As noted previously, an inorganic (mineral) acid is often used to accelerate the production time for solution-grade PAA. The catalyst most commonly used for wastewater disinfection products is sulfuric acid. And, while sulfuric acid is often used to drive the speed of reaction to equilibrium, it is not strictly required and is only used as an optional means of expediting production rates.

2.1.2 Stabilizers

Stabilizers are added to commercial products to avoid the occurrence of decomposition phenomena and preserve the manufactured composition. Typical stabilizers include compounds such as dipicolinic acid, alkali metal polyphosphates, or quinolone derivatives (U.S. Patent No. 3,442,937, 1969). The most common stabilizer is hydroxyethylidene diphosphate (HEDP), which is present at very low concentrations in PAA formulations. Although other stabilizers may be used, HEDP is used in most formulations for wastewater disinfection applications (according to a review of several PAA formulations that are used for toll blending purposes). The use of HEDP in PAA formulations for food contact applications is well documented in U.S. Food and Drug Administration food contact notifications, which include information on the environmental fate and ecotoxicity of HEDP.

2.2 On-Site Generation of Peracetic Acid

As an alternative to purchasing stabilized commercial-solution grade PAA, the chemical can be produced on-site. While there are currently no known wastewater treatment facilities in North America using on-site generation, the growing interest in use of PAA for disinfection, along with commoditization of the PAA market, may drive the interest toward potentially more economical methods of delivering PAA. There are several methods used for on-site generation. At least one system provider provides equipment to produce distilled PAA, which begins in a similar fashion as making solution-grade PAA, then uses continuous distillation to drive off both PAA and water, producing the final purified PAA product. This approach allows all reactants to be converted into PAA product. Another system provider uses an electrolytic process to generate PAA at concentrations up to 3.5% from concentrated feedstock chemicals, hydrogen peroxide, sulfuric acid, and acetic acid; although the final solution strength is low, the system can work at a range of temperatures up to 55 °C (131 °F).

Although there may be interest in use of on-site generation, most municipal wastewater treatment facilities would consider a commercially available U.S. EPA-registered PAA solution over on-site generation for several reasons. First, PAA produced on-site is typically only stable for 24 to 48 hours, and only when stored under refrigerated conditions (<5 °C [<41 °F]). Additionally, the simplicity of the chemical storage and feed systems required to feed solution-grade PAA is an attractive operational consideration. For these reasons, it is anticipated that bulk solution-grade PAA will dominate the wastewater disinfection market; thus, this document is focused on addressing application of PAA disinfection of wastewater using commercial solution-grade PAA.

3.0 U.S. ENVIRONMENAL PROTECTION AGENCY REGISTERED PRODUCTS AND MANUFACTURERS

As with any disinfectant, the PAA product must be registered with the appropriate federal agencies, such as U.S. EPA in the United States and Health Canada in Canada. In the United States, the product label must list the specific product formula and the statements of use. Before approving the product, the governing regulatory agency must review and accept efficacy data specific to the use conditions, including application rates and effluent residual concentrations, as well as relevant toxicity data (aquatic and ecotoxicity). Uses outside the approved label statements, either for different applications or at different application rates, are illegal under the Federal Insecticide, Fungicide, and Rodenticide Act (FIFRA) guidelines in the United States or Pest Management Regulatory Agency (PRMA) guidelines in Canada.

Subsequent to federal approval, each state (or provincial) regulatory agency can accept or decline use of the biocide within their jurisdiction. State or provincial agencies can set tighter application rates or residual concentrations than those approved on the federal product label, but they cannot add applications that extend beyond the stated label requirements.

There are several commercial PAA products that have been registered with U.S. EPA for use in wastewater disinfection. Table 2.2 summarizes the products that are commonly being used for the stated purpose at the time of this publication.

4.0 PERACETIC ACID SHIPPING REQUIREMENTS

Procedures for transporting PAA depend on the classification of PAA solutions that is defined by national and international regulations. Such classification is typically related to the PAA and hydrogen peroxide content of commercial products. Peracetic acid solutions are usually transported by road, rail, and sea using high-density polyethylene drums or stainless steel tanks approved by competent authorities (European Centre for Ecotoxicology and Toxicology of Chemicals, 2001). In the United States, PAA can only be shipped by truck, and not by rail. Transportation may impact the cost of PAA solutions because this cost can vary depending on distance

TABLE 2.2 Peracetic acid products with U.S. EPA registration for use in wastewater disinfection.

Manufacturer	Solvay	Evonik	PeroxyChem	BioSide	EnviroTech	EnviroTech
	Proxitane WW-12	Peraclean 15[a]	VigorOx WWT II	SaniDate 15[a]	Peragreen WW	Peragreen 22WW
U.S. EPA registration	68660-1	54289-4	6540/2-3	70299-26	63838-2	63838-20
Registration date	2013	2015	2008	2015	2015	2015
Application rate	0.5–10 ppm[b]	0.5–15 ppm	0.5–25 ppm	0.5–15 ppm	0.5–10 ppm	0.5–10 ppm
Allowable residual	1.0 ppm	1.0 ppm	1.0 ppm; if DF[c] > 12, then 0.09 × DF	1.0 ppm	1.0 ppm	1.0 ppm
Chemical composition						
Peracetic acid	12%	15%	15%	15%	14.8–15.7%	21.5%
Hydrogen peroxide	18.5%	22%	23%	10%	21.5–22.5%	5%
Acetic acid	20%	15–20%	16%	38–41%	15.8–16.6%	40–50%
Water	Balance	Balance	45%	Balance	Balance	Balance
Freezing point	−30 °C (−22 °F)	−50 °C (−58 °F)	−49 °C (−56 °F)	No data available	−12 °C (10 °F)	−17 °C (0 °F)

[a]Label not for use in California
[b]ppm = parts per million
[c]DF = dilution factor

to manufacturing facilities. Thus, pricing is often customer- and location-specific and dependent on quantities needed. Fluctuations in transportation costs often impact PAA costs. Chapter 12 includes additional information on cost.

5.0 HANDLING AND STORAGE REQUIREMENTS FOR PERACETIC ACID

Peracetic acid should be stored in a well-ventilated place, with average temperatures not exceeding 30 °C (86 °F) during the storage period. In North America, PAA for wastewater disinfection is commercially available in 50-gal (190 L) drums, 300-gal (1135 L) totes, and in bulk. When considering storage of drums and totes, PAA must be kept in original containers and, preferably, away from sunlight (Kitis, 2004). When PAA is procured in bulk, storage tanks should be fabricated of compatible materials such as high-density polyethylene (HDPE) grades and properly passivated stainless steel 304L or 316L. However, because PAA can embrittle plastics and elastomers, and may leach metal ions from stainless steel with extended contact times, PAA should not be stored in the same container indefinitely. For example, PAA supplied in intermediate bulk container (IBC) totes should be consumed before the tote expiration date, which is set by the manufacturer and displayed on the tote face plate. The mechanical integrity of the tote may be compromised if PAA remains in the tote beyond the tote expiration date, and special hazardous material procedures may be required to remove PAA from expired totes.

On the other hand, PAA stored in stainless steel tanks will not affect the mechanical integrity of the steel; however, the exposure to higher concentrations of metal ions that may leach over time could reduce the SADT and trigger PAA decomposition at normal storage temperatures (below 30 °C [86 °F]) (European Centre for Ecotoxicology and Toxicology of Chemicals, 2001). Proper storage of product must include preventing the contamination of PAA with incompatible materials such as reducing agents, organic compounds, and carbon steel objects. Table 2.3 lists materials of compatibility for PAA.

In the United States, water resource recovery facilities (WRRFs) are exempt from Department of Homeland Security (DHS) requirements related to Chemical Facility Anti-Terrorism Standards (CFATS) and are only regulated by U.S. EPA, not DHS, as defined in Section 212 of the Federal Water Pollution Control Act 33 U.S.C § 1292. As a result, the use of PAA is excluded from the U.S. EPA Risk Management Plan (RMP) requirements, and WRRFs are only required to adhere to the general duty clause of CFATS.

In reference to peracetic acid mixtures and the U.S. EPA RMP, according to 40 C.F.R. 68.115(a)(b)(1), "... *if the concentration of the regulated substance in the mixture is one percent or greater by weight, but the owner or operator can demonstrate that the partial pressure of the regulated substance in the mixture (solution) under handling or storage conditions in any portion of the process is less than 10 millimeters of mercury (mm Hg), the amount of the substance in the mixture in that portion of the process need not be considered when determining whether more than a threshold quantity is present at a stationary source*".

The products that are U.S. EPA-registered for wastewater disinfection are mixtures containing 12 to 22%, by weight, of PAA. The PAA vapor pressure in these mixtures is substantially less than the indicated threshold value. For example, the PeroxyChem product, VigorOx II, has an estimated vapor pressure of 3.12 mm Hg at 50 °C (122 °F), and 0.361 to 0.681 mm Hg at 55 to 73 °F (13 to 23 °C). This estimate was calculated using tabulated data and Raoult's Law. Similar calculations can be made for other commercially available products and, based on this information, the PAA in the storage and handling system used at wastewater treatment facilities does not need to be considered toward the threshold quantity. Thus, products containing 22% by weight of PAA or less would satisfy the stationary threshold and may be used at a WRRF without an RMP.

5.1 Safety

Commercial bulk solution-strength PAA is classified as a hazardous material under the Occupational Safety and Health Agency Hazard Communication Standard (29 C.F.R. 1910.1200). The following hazard statements are required for labeling these solutions with respect to handling dangers (Figure 2.3 also shows several of these hazard symbols):

- Causes serious eye damage
- Causes severe skin burns and eye damage
- Heating may cause a fire

FIGURE 2.3 Hazard symbols required on labels of commercial solution-grade PAA products (graphic courtesy of PeroxyChem).

- Harmful if swallowed
- Harmful in contact with skin
- Harmful if inhaled
- May cause respiratory irritation

Site-specific safety plans should be prepared for PAA installations, as further described in Chapter 9. Chapter 9 also describes design considerations for PAA systems, such as venting to prevent pressure buildup.

5.1.1 Precautionary and Exposure Procedures

Personal protective equipment (PPE) should also be used in handling PAA. The product-specific SDS should be followed; however, a, number of general measures can be implemented to prevent hazards associated with use of PAA. The following general prevention measures should be taken when handling PAA solutions for wastewater disinfection:

- Wear appropriate PPE, including gloves, clothing, and eye and face protection, such as rubber or neoprene footwear; impervious clothing materials such as rubber, neoprene, nitrile, or polyvinyl chloride; and liquid-proof rubber or neoprene gloves.
- Store PAA away from combustible materials.
- Keep PAA away from heat/sparks/open flames/hot surfaces, and do not smoke near PAA storage or feed facilities.
- Keep PAA only in its original container.
- Do not breathe PAA fumes, gas, mist, vapors, or spray.

In the event PAA is spilled, there are several response actions that can be taken to minimize damage depending on the affected area. If PAA is spilled on clothing or equipment, immediately remove contaminated clothing, rinse exposed skin with water, and wash contaminated clothing before reuse. Because the liquid and mist are corrosive (i.e., causes burns), direct contact could cause irreversible damage to the eyes, including blindness and/or irreversible destruction of skin tissue. Peracetic acid vapor or mist will irritate the nose, throat, and lungs, but will typically subside when exposure ceases. The severity of exposure effects depends on the chemical concentration and exposure. If exposure occurs, the following response and first aid measures should be taken:

- Eye contact—remove contact lenses, if present, and rinse with water for at least 15 minutes; continue rinsing and immediately call a poison center or physician.

- Skin contact—wash off affected area immediately with soap and plenty of water while removing all contaminated clothes and shoes. Call a physician immediately.
- Inhalation—move the person to fresh air and keep comfortable; if breathing difficulty or discomfort occurs and persists, obtain medical attention immediately.
- Ingestion—rinse out the mouth with fresh water and drink plenty of water, do NOT induce vomiting, and call a physician or poison control center immediately.

5.1.2 Flammability and Firefighting Considerations

Although PAA is not flammable, it undergoes rapid decomposition under fire conditions and releases oxygen, which intensifies the fire. It is, therefore, important to store PAA away from flammable materials such as wood pallets. In case of fire, use water as the extinguishing media. Cool containers by flooding quantities of water on them until well after the fire is out. Note that chemical-type extinguishers are not effective with PAA. Fight fire from maximum distance or use unmanned hose holders or monitor nozzles. Wear a self-contained breathing apparatus and protective suit. Always refer to the product-specific SDS.

5.1.3 Spill Response Considerations

In case of loss of containment, first responders should wear appropriate PPE including chemically resistant goggles, gloves, boots, and outer garments. The spill should be isolated from traffic areas by posting hazard signs and cordoning off the area with caution/hazard tape. All sources of ignition (spark or flame) should be removed when possible, including turning off electrical equipment and removing combustible material such as wooden pallets that may catch fire when in contact with PAA. When possible, the spill should be contained to prevent contamination of soil, stormwater drains, sewer drains, and groundwater by deploying containment berms or dikes made of chemically resistant materials. Finally, if proper containment has been achieved, water can be used to dilute PAA and facilitate neutralization and removal by hazardous material handling specialists. Peracetic acid manufacturers provide emergency contacts to assist first responders during spill containment and removal.

5.2 Materials of Compatibility Overview

Table 2.3 presents a list of materials and their compatibility with PAA. The information was compiled from sources including manufacturers, suppliers, and end users.

TABLE 2.3 Peracetic acid materials of compatibility.

Common name	Other name	Type	Compatibility[a]	Chemical resistance source
Butyl[c]	Rubber, polyisobutylene	Elastomer	A	CR1, CR3
Fluorosilicone	FSR, silicone rubber	Elastomer	A	CR1, CR3, CR7
Hypalon	Chlorosulfonated polyethylene, CSPE	Elastomer	A	CR1, CR3
Kalrez	Perfluoroelastomer, FFKM	Elastomer	A C	CR1 CR7
Neoprene[b, c]	Polychloroprene	Elastomer	A D	CR1, CR3 CR7
SBR	Styrene-butadiene	Elastomer	A	CR1, CR3
EPDM[b]	Ethylene propylene diene	Elastomer	A B C	CR1, CR2, CR3 CR4, CR5, CR7 CR9
Silicone[b]	Silicon rubber, polysiloxane	Elastomer	B C	CR1, CR2, CR3 CR7
Viton[b]	Fluoroelastomer polymer	Elastomer	C A	CR1, CR3 CR2, CR5, CR7
Nitrile	Buna-N, perbunan, acrylonitrile, butadiene rubber, NBR	Elastomer	C	CR1, CR2, CR3, CR7
304L stainless steel (passivated)		Metal	B	CR9
316L stainless steel (passivated)		Metal	A	CR6, CR9
Aluminum[c, d]		Metal	B	CR6
Ceramic		Other	A	CR6
Glass		Other	Good	
Kynar[2]	PVDF, polyvinylidede fluoride	Polymer (plastic)	N/A A	CR2 CR4

(*continues*)

TABLE 2.3 Peracetic acid materials of compatibility (*Continued*).

Common name	Other name	Type	Compatibility[a]	Chemical resistance source
PEEK	Polyether ether ketone	Polymer (plastic)	A	CR6
PTFE	Teflon, polytetrafluoroethylene	Polymer (plastic)	A	CR4, CR5, CR6, CR8
HDPE	High-density polyethylene	Polymer (plastic)	A	CR8
Acrylate	MAA, polyacrylate, methyl acrylate	Polymer (plastic)	D	CR1, CR3
CPVC	Chlorinated polyvinyl chloride	Polymer (plastic)	D	CR4, CR5
LDPE	Low-density polyethylene	Polymer (plastic)	D	CR8
PP[b]	Polypropylene	Polymer (plastic)	A D	CR2 (15%) CR4, CR5 (40%)
PVC	Polyvinyl chloride	Polymer (plastic)	D	CR4, CR5, CR6

Notes

[a]Compatibility ratings have been compiled from various sources, some with differing compatibility notations/ranking. This table presents a summary of chemical resistance, but the original sources should be referenced to appreciate the differences between ratings such as excellent, fair, good, doubtful, etc.

[b]Conflicting compatibility ratings and their references have been noted in some instances; refer to the source references for details.

[c]Manufacturer input recommends short-term exposure for this material.

[d]Not recommended for piping or tanks.

Key

Compatibility: A = Excellent/Satisfactory, B = Good/Fair, C = Fair/Doubtful, D = Not recommended/ Unsatisfactory, N/A = Not available (insufficient data)

Chemical resistance references:

CR1: http://mykin.com/rubber-chemical-resistance-chart-5

CR2: https://www.verderflex.com/fileadmin/files/verderflex/documents/Verderflex_Chemical_Compatibility_Guide/Verderflex_Chemical_Compatibility_Guide.pdf

CR3: http://www.sealanddesign.com/page/fluid-compatibility

CR4: https://www.haywardflowcontrol.com/assets/documents/flowcontrol/pdf/ChemChart.pdf

CR5: http://www.lutzjescoamerica.com/fileadmin/data/data_usa/documents/Chemical_Resistance_Chart.pdf

CR6: http://www.trimec-europe.com/assets/useful/ChemicalCompatibility.pdf

CR7: https://www.industrialspec.com/images/files/elastomers-chemical-compatibility-chart-from-ism.pdf

CR8: http://envirotech.com/wp-content/uploads/2015/12/PAA-101.pdf

CR9: http://www.peroxychem.com/media/121257/feb2015_safeuseandhandling.pdf

5.2.1 Long-Term Exposure

There are no readily available data in the literature regarding effects of long-term exposure to PAA or long-term exposure limits. Most of the published data relate to acute and short-term exposure effects. Peracetic acid is a strong oxidant and, as such, long-term exposure could lead to irritation of the respiratory tract, skin, and eyes. Facilities using PAA should consider implementation of engineering or procedural controls that prevent long-term or continuous exposure of personnel to PAA vapors in the air.

5.2.2 Short-Term Exposure

The short-term exposure limit for PAA vapors in the air is 0.4 ppm over a rolling 15-minute period. This limit was established by the American Conference of Governmental Hygienists in 2014 based on toxicity data and is meant to prevent personnel exposure to PAA concentrations in the air that can lead to irritation of the eyes, skin, and respiratory tract. Facilities using PAA should evaluate implementation of engineering and procedural controls that minimize personnel exposure to PAA vapors and require the use of the appropriate PPE such as respirators, chemical-resistant suits, and eye protection.

5.3 Other Safety Considerations

Although PAA is a strong oxidant with specific storage and handling requirements, the fact that it is transported and stored as a liquid reduces safety concerns associated with loss of containment during transportation or storage resulting from accidents or terrorist attacks, which are one of the main drivers for replacement of chlorine gas.

6.0 REFERENCES

Block, S. S. (2001). *Disinfection, sterilization, and preservation* (5th ed.). Lippincott Williams & Wilkins. Retrieved from https://trove.nla.gov.au/work/6733270

Domínguez Henao, L., Delli Compagni, R., Turolla, A., & Antonelli, M. (2018). Influence of inorganic and organic compounds on the decay of peracetic acid in wastewater disinfection. *Chemical Engineering Journal,* 337. doi:10.1016/j.cej.2017.12.074

Drimus, J., & Matasa, C. (1966). Spontaneous decomposition of peracetic acid. *Bulletin de la Societe Chimique de France*, 12, 253–255.

European Centre for Ecotoxicology and Toxicology of Chemicals. (2001). Joint Assessment of Commodity Chemicals (JACC) programme. JACC No. 40—Peracetic Acid (CAS No. 79-21-0) and its Equilibrium Solutions. Brussels, Belgium.

Gunter, L., Heinrich, R., & Kurt, S. (1969). *U.S. Patent No. 3,442,937. Process for stabilizing solutions of aliphatic percarboxylic acids*. Washington, DC: U.S. Patent and Trademark Office.

Kitis, M. (2004). Disinfection of wastewater with peracetic acid: A review. *Environment International*, 30, 47–55. doi:10.1016/S0160-4120(03)00147-8

Kunigk, L., Gomes, D. R., Forte, F., Vidal, K. P., Gomes, L. F., & Sousa, P. F. (2001). The influence of temperature on the decomposition kinetics of peracetic acid in solutions. *Brazilian Journal of Chemical Engineering,* 18, 217–220. doi:10.1590/S0104-66322001000200009

Kunigk, L., Silva, S. M., & Jurkiewicz, C. H. (2012). The influence of temperature and organic matter on the decomposition kinetics of peracetic acid in aqueous solutions. *Latin American Applied Research Journal,* 42, 291–297. Retrieved from http://www.scielo.org.ar/pdf/laar/v42n3/v42n3a11.pdf

O'Sullivan, D. W., Lee, M., Noone, B. C., & Heikes, B. G. (1996). Henry's law constant determinations for hydrogen peroxide, methyl hydroperoxide, hydroxymethyl hydroperoxide, ethyl hydroperoxide, and peroxyacetic acid. *The Journal of Physical Chemistry,* 100, 3241–3247. doi:10.1021/jp951168n

Pedersen, L. F., Meinelt, T., & Straus, D. L. (2013). Peracetic acid degradation in freshwater aquaculture systems and possible practical implications. *Aquacultural Engineering*, 53, 65–71. doi:10.1016/j.aquaeng.2012.11.011

Swern, D. (1970). *Organic peroxides*. Retrieved from https://books.google.it/books/about/Organic_peroxides.html?id=oV1it0e3-mwC&redir_esc=y

U.S. Environmental Protection Agency. (1976). *Disinfection of wastewater—task force report* (EPA-430/9-75-013). Washington, DC: US Environmental Protection Agency.

U.S. Environmental Protection Agency. (2015). *Update of human health ambient water quality criteria: Cyanide 57-12-5*. Washington, DC: US Environmental Protection Agency.

Yuan, Z. (1997) Kinetics of peracetic acid decomposition: Part I: Spontaneous decomposition at typical pulp bleaching conditions. *Canadian Journal of Chemical Engineering,* 75, 37–41. doi:10.1007/s00761-002-0289-9

Yuan, Z., Ni, Y., & Van Heiningen, R. P. (1997). Kinetics of the peracetic acid decomposition: Part II: pH effect and alkaline hydrolysis; *Canadian Journal of Chemical Engineering,* 75, 42–47. doi:10.1002/cjce.5450750109

Zhao, X., Cheng, K., Hao, J., & Liu, D. (2008). Preparation of peracetic acid from hydrogen peroxide: Part II: Kinetics for spontaneous decomposition of peracetic acid in the liquid phase; *Journal of Molecular Catalysis A: Chemical,* 284, 58–68. doi:10.1016/j.molcata.2008.01.003

Zhao, X., Zhang, T., Zhou, Y., & Liu, D. (2007). Preparation of peracetic acid from hydrogen peroxide: Part I: Kinetics for peracetic acid synthesis and hydrolysis. *Journal of Molecular Catalysis A: Chemical,* 271, 246–252. doi:10.1016/J.MOLCATA.2007.03.012

Zhao, X., Zhang, T., Zhou, Y., & Liu, D. (2008). Preparation of peracetic acid from acetic acid and hydrogen peroxide: Experimentation and modeling. *The Chinese Journal of Process Engineering*, *8*(1), 35–41.

3

Peracetic Acid Disinfection Kinetics

Josh Goldman, Siva Sarathy, Kyriakos Manoli, Roberta Maffettone, Andrea Turolla, Manuela Antonelli, Laura Dominguez Henao, Joseph Jacangelo, & Domenico Santoro

1.0 PERACETIC ACID MECHANISMS OF MICROBIAL INACTIVATION

This chapter provides an overview of the current state of knowledge about how peracetic acid (PAA) works. First, the most current research on mechanisms of action of PAA toward bacteria and viruses is explained, and the kinetics of those reactions are described. Next, a review of combined disinfection using PAA and ultraviolet (UV) irradiation is presented that includes an explanation of the PAA photochemistry; a discussion about the possible synergies between the two technologies; a proposed modeling technique to quantify combined disinfection; and, finally, an economic analysis to determine when combined disinfection is cost-effective. Next, the kinetics of PAA decomposition and pathogen inactivation in wastewater is explained, and modeling techniques are described to quantify and predict inactivation. Following this is a detailed discussion about how commonly occurring water quality constituents in wastewater affect PAA decomposition and inactivation kinetics. Lastly, PAA dosing control strategies are explained and compared.

Among the compounds present in a PAA equilibrium quaternary mixture (i.e., acetic acid [CH_3COOH], hydrogen peroxide [H_2O_2], PAA [CH_3COOOH], and water [H_2O]), PAA itself has been shown to have the greatest antimicrobial properties compared to hydrogen peroxide and acetic acid (Kitis, 2004; Wagner, Brumelis, & Gehr, 2002). The mode of microbial inactivation of PAA for disinfecting wastewater has received increasing attention over the past two decades. Different mechanisms based on different types and species or organisms have been advanced in the literature.

In investigating the mode of action of PAA, it is important to understand the speciation of PAA as a function of pH. Peracetic acid dissociates to peracetate anion (CH_3COOO^-), which has a pK_a of 8.2. At pH less than 8, the undissociated acid (CH_3COOOH) is the primary molecule and has greater antimicrobial activity than the dissociated peracetate anion, which is prevalent at pH greater than 8 (Koivunen & Heinonen-Tanski, 2005b; Liberti & Notarnicola, 1999). The high biocidal efficacy of PAA at pH less than 8 may be explained by the higher redox potential of PAA at lower pH (i.e., $E^0 = 1.748$ V versus the standard hydrogen electrode [SHE] at pH = 0, and $E^0 = 1.005$ V vs. the SHE at pH = 14) (Dunkin et al., 2018; Zhang, Brown, & Hu, 2018).

In addition to the previously discussed PAA species, the disinfection (e.g., inactivation of microbes) and oxidation (e.g., degradation of micropollutants) properties of PAA in wastewater treatment can also be attributed to the formation of reactive oxygen species (ROS) such as hydroxyl radical ($HO\cdot$), superoxide anion ($O_2^{-\cdot}$), hydroperoxyl radical ($HO_2^{\cdot}$), and peracyl radicals

($CH_3COO\cdot$). These reactive species may contribute to the inactivation of bacterial cells (Block, 2001). The formation of the aforementioned ROS during PAA disinfection may occur through multiple reactions (Equations 3.1 to 3.7). The reaction described in Equation 3.1 represents the formation of the acyloxy and hydroxyl radicals via homolysis of PAA. This reaction may be initiated by activation of PAA by a transition metal catalyst or through UV radiation (Luukkonen & Pehkonen, 2017; Rokhina et al., 2010).

$$CH_3COOOH \rightarrow CH_3COO\cdot + HO\cdot \tag{3.1}$$

$$CH_3COOOH + HO\cdot \rightarrow CH_3CO\cdot + O_2 + H_2O \tag{3.2}$$

$$CH_3COOOH + HO\cdot \rightarrow CH_3COOO\cdot + H_2O \tag{3.3}$$

$$CH_3COO\cdot \rightarrow \cdot CH_3 + CO_2 \tag{3.4}$$

$$2CH_3COO\cdot \rightarrow 2\cdot CH_3 + 2CO + O_2 \tag{3.5}$$

$$\cdot CH_3 + O_2 \rightarrow CH_3OO\cdot \tag{3.6}$$

$$CH_3COO\cdot + HO\cdot \rightarrow CH_3COOOH \tag{3.7}$$

Disinfection by PAA occurs according to reactions with the bacterial wall, the cytoplasmic membrane, or the metabolisms of the cell (Figure 3.1). These include (a) disruption of cell wall and chemiosmotic function; (b) disruption

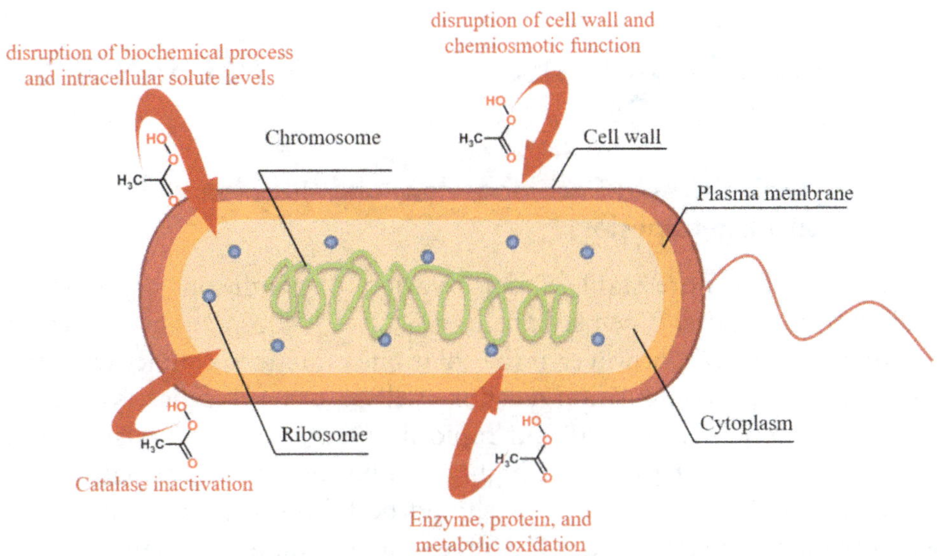

FIGURE 3.1 Illustration of disruption of cellular targets by PAA.

of biochemical process and intracellular solute levels; (c) enzyme, protein, and metabolic oxidation; and (d) reduction of catalase activity.

Reactions of PAA with sulphydryl (-SH) groups, the disulfide (-S-S) bonds, and double bonds within enzymes, proteins, and other metabolites contained in the cell membrane wall disrupt the chemiosmotic function of the lipoprotein cytoplasmic membrane and increase cell wall permeability (Baldry & Fraser, 1988; Lefevre, Audic, & Ferrand, 1992; Liberti & Notarnicola, 1999). Peracetic acid may also penetrate intracellular enzymes and oxidize essential enzymes (e.g., the peroxidase), causing the weakening of biochemical routes and the disruption of the intracellular solute levels (Fraser, Godfree, & Jones, 1984). The ovicidal properties of PAA are probably related to its protein denaturation action, exemplified by its use as a sporocide (Block, 2001). The nucleotide bases of DNA molecules may also be affected by PAA, disrupting cell replication. As opposed to the oxidizing action of hydrogen peroxide, which is inhibited through the reduction of free hydroxyl radicals by catalase enzymes found in all living organisms, PAA exerts damage through inner diffusion (Block, 2001).

Despite the progress in investigating the modes of action of PAA in microbial inactivation, these modes are not thoroughly understood. Peracetic acid inactivation in municipal wastewater is complex because it results from combined pathways and simultaneous routes and modes of action. Moreover, organic and inorganic constituents in municipal wastewater may promote or inhibit various reactions Thus, future research is necessary to further elucidate the process and role played by multiple species potentially involved in the inactivation of bacteria, viruses, and protozoa by PAA.

2.0 PERACETIC ACID DECOMPOSITION MECHANISMS AND KINETICS

2.1 Hydrolysis, Auto-Decomposition, and Reaction with Non-Target Species

Peracetic acid is more stable under a variety of conditions compared to sodium hypochlorite. When kept below 13 °C (55 °F), spontaneous decomposition is negligible; however, if the solution comes into contact with metals or organics, decomposition reactions will occur (Jurkiewicz, Kunigk, & Silva, 2012; Zhao et al., 2008). If rapid decomposition occurs as a result of these reactions, the result is highly exothermic (Vianello, Salzano, & Maschio, 2018); thus, procedures should be followed to prevent contact with organics or metals, and safety precautions should be taken to prepare for these types of reactions in case they do occur.

Greenspan & Gall (1956) found that a 40% solution of PAA at room temperature decayed by 1 to 2% per month. Dilute solutions of PAA (below 1%) are unstable (Kunigk & Almeida, 2001). PeroxyChem, a major PAA manufacturer and supplier, reports that their 15% solution is formulated to maintain equilibrium for 1 year when stored at temperatures below 29 °C (84 °F). The rate of the decay (day^{-1}) is related to temperature by the Arrhenius equation (Equation 3.8), where A is the pre-exponential factor, E is the activation energy, R is the universal gas constant, and T is temperature in °K. Table 3.1 shows the approximate shelf life of a 15% PAA solution at three temperatures (PeroxyChem, 2014).

$$k = Ae^{\frac{-E}{RT}} \tag{3.8}$$

When dosed in typical municipal wastewater, PAA is generally consumed in three competing reactions (Equations 3.9, 3.10, and 3.11). Spontaneous decomposition occurs when PAA degrades into acetic acid and oxygen. In the hydrolysis reaction, it reacts with water to form acetic acid and hydrogen peroxide. Hydrogen peroxide quickly decomposes into water and oxygen. At the same time, PAA reacts with any transition metals that are present to form acetic acid, oxygen, and other byproducts (PeroxyChem, 2015).

$$2CH_3CO_3H \rightarrow 2CH_3CO_2H + O_2 \tag{3.9}$$

$$CH_3CO_3H + H_2O \rightarrow CH_3CO_2H + H_2O_2 \tag{3.10}$$

$$CH_3CO_3H + M^+ \rightarrow CH_3CO_2H + O_2 + \text{other products} \tag{3.11}$$

Numerous studies have shown that the decomposition of PAA occurs in two distinct and independent phases and can be modeled by a modified first order kinetic equation (Equation 3.10) (Dell'Erba et al., 2004). The first phase is an instantaneous reduction that occurs as soon as the PAA contacts the wastewater. This phase of decomposition is mainly controlled by water quality and is typically associated with oxidation of organic matter (Domínguez Henao, Delli Compagni, Turolla, & Antonelli, 2018). Recent

TABLE 3.1 Approximate shelf life of 15% PAA formulations.

Temperature	Approximate shelf life
<30 °C (<86 °F)	1 year
<37.8 °C (<100 °F)	4 months
<43.3 °C (<110 °F)	1 month

bench-scale studies designed by Domínguez et al. to quantify PAA demand and decay concluded that initial PAA demand was primarily a result of the reaction of disinfectant with organic material, and that this type of consumption occurred within the first 5 minutes after initially dosing the PAA. Bench studies concluded that casein and peptone, in particular, were the predominant cause of most of the initial demand because hemolytic fission of PAA generates peroxyl and hydroxyl radicals that are highly reactive with proteins (Domínguez Henao, Delli Compagni, Turolla, & Antonelli, 2018). Initial demand may also be related to particles and microorganisms (Antonelli et al., 2013; Falsanisi et al., 2006).

The second phase of decomposition is characterized by a first order decay coefficient and is associated with hydrolysis and transition metal-catalyzed decomposition (Dell'Erba et al., 2004). Exponential decay has been shown to be associated with inorganics such as reduced iron and orthophosphate (Domínguez Henao, Delli Compagni, Turolla, & Antonelli, 2018). Iron consumes PAA directly by catalysis, whereas orthophosphate acts as chelating compound toward iron, reducing the effect of iron (Domínguez Henao, Delli Compagni, Turolla, & Antonelli, 2018; Liu et al., 2014). Liu et al. (2014) reported that decay occurs more rapidly in waters with high salinity, such as seawater. Results from this study indicated that higher ratios of sodium and potassium to calcium and magnesium significantly increased the PAA degradation rate in 1 and 3% saltwater solution.

The magnitude of both the instantaneous demand (D) and the first order decay coefficient, k, is dependent on initial dose, PAA formulation (%PAA), pH, salinity, and hardness (Luukkonen & Pehkonen, 2017). When PAA is added to demand-less water, the concentration decreases 25 to 30% in 1 hour, at a rate consistent with first order kinetics. Generally, PAA demand varies more than decay over various water qualities (Dell'Erba et al., 2004); however, Falsanisi et al. (2006) observed a 10-times greater decay rate in primary effluent versus secondary effluent.

2.2 The Instantaneous Demand Equation

Peracetic acid residual concentrations can be predicted by fitting parameters k and D in the standard decomposition equation (Equation 3.12) (Antonelli et al., 2013; Haas & Finch, 2001) from an experimental dataset, where C is the PAA concentration at time t, C_0 is the initial PAA dose, D is the instantaneous demand, and k is the decay coefficient. This is performed using an iterative solver and minimizing the sum square difference between the measured and calculated residual.

$$C = (C_0 - D) \times e^{-kt} \tag{3.12}$$

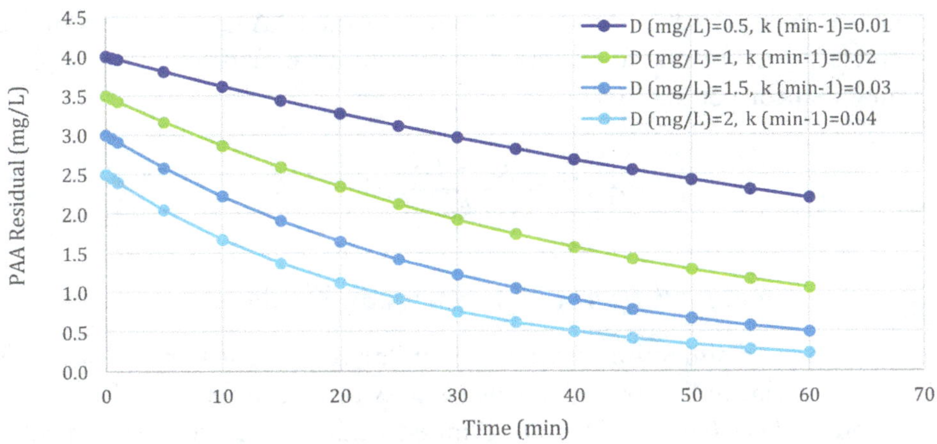

FIGURE 3.2 Typical PAA decay curves for an initial dose (C_o) of 4.5 mg/L at varying demands and decay rates.

The goodness of fit will be affected by changes in water quality. In datasets in which there are large variations in water quality, it may be necessary to determine more than one set of fitting parameters by separating the dataset into subsections that relate to specific water quality conditions, such as changes in inorganic loading. Another option is to model one or both of the fitting parameters as a function of water quality variables such as total suspended solids (TSS), color, or chemical oxygen demand (COD). Table 3.2 shows typical demand and decay values for PAA in wastewater.

The efficacy of PAA for the inactivation of microorganisms is most often considered in terms of a CT concept, which is based on both time of exposure and residual concentration. When calculating the PAA CT, it is important to use the integral method instead of simply multiplying the residual by the exposure time (or hydraulic residence time, HRT, in reactor); because this accounts for the exponential decay of PAA. PAA CT can be visualized as the area under the decay curve, and it is calculated by integrating the decay equation from time zero to time *t*. Equation 3.13 shows the analytical solution to the decay equation, where C_0 is the initial PAA dose, D is the instantaneous demand, and k is the decay constant. If conventional methods are used to determine the CT, a significant amount of error will

TABLE 3.2 Typical demand decay rates in wastewater.

Parameter	Typical values
D (demand) mg/L	0.5–3.0
K (decay) min-1	0.01–0.06

occur (Rossi et al., 2007). Figure 3.3 compares the PAA CT calculated using the integral method to values based simply on dose or residual, multiplied by time (Santoro et al., 2015).

$$PAA\ CT = \frac{(C_0 - D)}{k} \times (1 - e^{-kt}) \tag{3.13}$$

2.3 Peracetic Acid Microbial Inactivation Kinetics

Peracetic acid disinfection efficacy is similar to that of hypochlorite for typical bacterial indicator organisms such as total and fecal coliform bacteria and *Escherichia. coli (E. coli)*. Limited studies have been performed to measure PAA disinfection efficacy toward viruses and phages, but a recent bench-scale study compared the disinfection efficacy of PAA to that of monochloramine toward the bacteriophage MS2 and marine norovirus (MNV). The study concluded that the efficacy of PAA toward MS2 was generally comparable to that of monochloramine as evidenced by comparable reductions in infectivity at similar applied contact time values. The same study indicated that PAA is slightly less effective than monochloramine at inactivating MNV and that MNV is generally less resistant than MS2 to both PAA and monochloramine (Dunkin et al., 2017). At typical contact times applied for wastewater that are effective for inactivation of bacterial targets, less than 1-log removal MS2 would be expected (Dunkin et al., 2017).

Although UV irradiation is generally considered a more effective disinfection technology than PAA, UV irradiation is more susceptible to changes in water quality and typically has less flexibility in dosing capabilities than PAA systems. Peracetic acid also has similar disinfection efficacy to chlorine dioxide, but the latter is usually more expensive to use because it needs to be generated in situ. A significant advantage of PAA over chlorine is that it does not generate regulated disinfection byproducts (Luukkonen & Pehkonen, 2017). Table 3.3 presents a literature review of PAA disinfection efficacy for several common organisms for different wastewater matrices (Luukkonen & Pehkonen, 2017).

Several mechanisms have been proposed to explain the mechanism of bacterial inactivation by PAA, although there is not yet a definitive explanation. The disinfectant's mode of action may be similar to other peroxides and oxidizing agents (Block, 2001). One study suggests that it is related to the oxidation of sulfhydryl (-SH) and sulfur (S-S) bonds in enzymes by "active oxygen" (Lefevre et al., 1992). Another proposed mechanism is the rupture of cell walls (Baldry & French, 1989). Peracetic acid may also oxidize enzymes that affect the vital biochemical pathways of microorganisms, and it may also transport directly across cell membranes (Fraser et al., 1984).

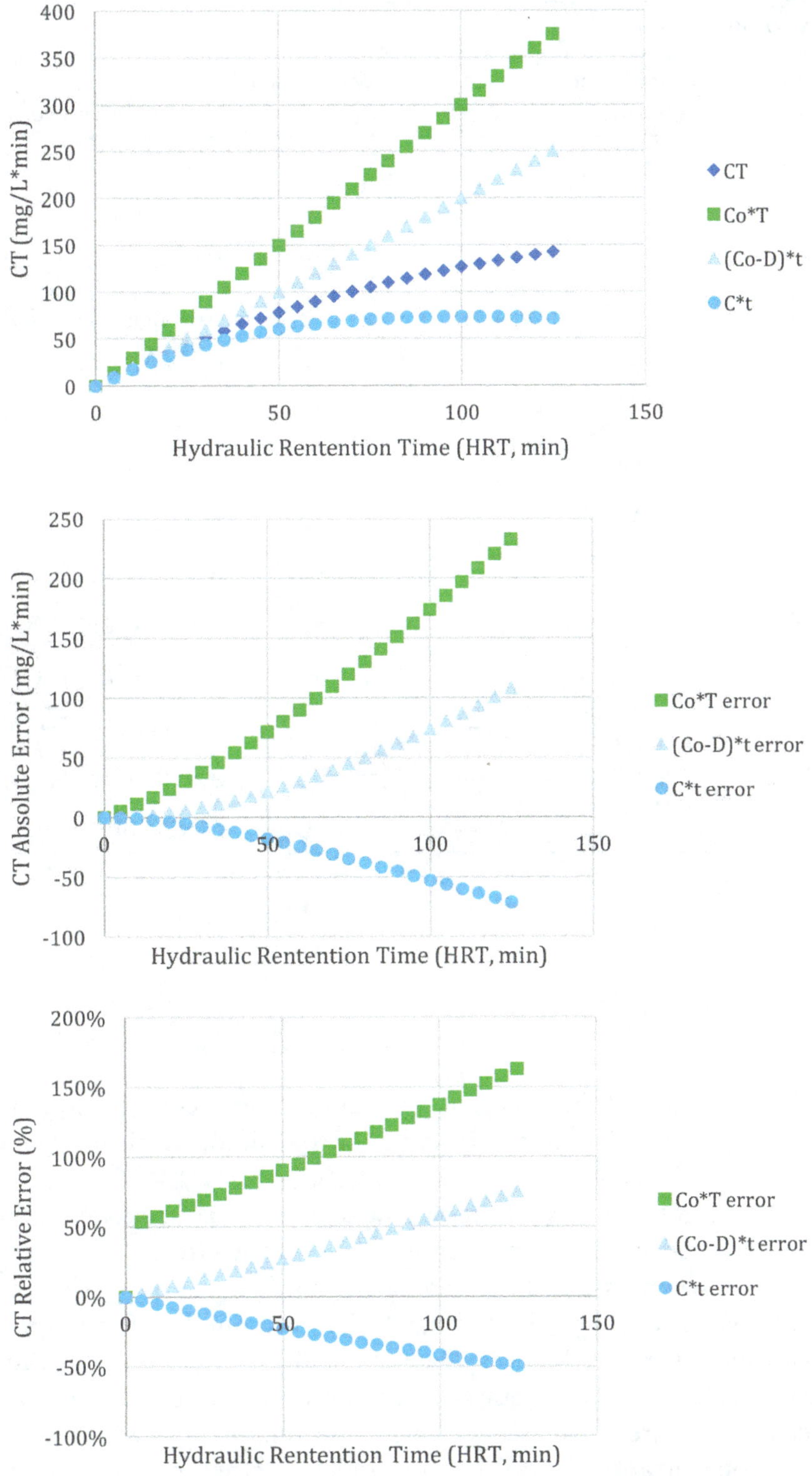

FIGURE 3.3 Error associated with using non-integrated CT calculation methods.

TABLE 3.3 Review of PAA disinfection efficacy.

Organism[a,b,c]	Dose (mg/L)	Contact time (min)	Log inactivation	Wastewater matrix[d,e]	Reference
EC	1.5	60	—	TE	Luukkonen et al., 2015
EC	2	55	—	SE	Rossi et al., 2007
EC	1–8	30	—	SE	Santoro et al., 2007
EC	5–10	35–50	4	SE	Antonelli et al., 2013
EC	4	10	2	SE	Santoro et al., 2007
EC	1.5	60	—	TE	Luukkonen et al., 2015
EC	2–25	6–54	—	TE	Azzellino et al., 2011
EC	2–25	6–54	—	TE	Mezzanotte et al., 2007
FC	5–15	55	—	SE	Rossi et al., 2007
FC	4.3	45	2	SE	Santoro et al., 2007
FC	1–8	30	—	SE	Santoro et al., 2007
FC	1–2	55	—	SE	Rossi et al., 2007
TC	2–25	6–54	—	TE	Azzellino et al., 2011
TC	2–25	6–54	—	TE	Mezzanotte et al., 2007

[a]EC = *E. coli*
[b]FC = fecal coliform
[c]TC = total coliform
[d]SE = secondary effluent
[e]TE = tertiary effluent

In addition, hydroxyl radicals generated during the decomposition process may directly participate in the inactivation of microorganisms (Bianchini et al., 2002; Lubello, Caretti, & Gori, 2002).

Disinfection efficacy is affected by wastewater characteristics, although some of these dependencies are more related to the effect on PAA decomposition rates, which are discussed earlier in this section. These include suspended solids and organic matter, which affect initial demand; transition metals, which increase the decay rate; and salinity, which abates radicals. Temperature affects the kinetics of both decomposition and inactivation. Higher temperatures increases the rate of decomposition, which decreases effective concentration (and CT) more quickly; however, higher temperatures also increase the reaction rate with microorganism, which in turn increases disinfection efficacy. Disinfection efficacy is greater at lower pH because the non-dissociated form of PAA is more active. Above pH 9, there is marked decreases in disinfection efficacy (Luukkonen et al., 2014).

Inactivation models relate the PAA concentration and contact time to the log inactivation of target microorganisms. Most validated inactivation models are related to the integral of the generalized inactivation rate (Equation 3.14),

$$\frac{dN}{dt} = -k\ C^n m N^x t^{m-1} \tag{3.14}$$

where

n, m, and x are dimension fitting parameters;
k' = rate of microbial decay (CFU/100mL/min);
N = microbial density (CFU/100mL);
C = disinfectant concentration (mg/L);
t = contact time (min).

Table 3.4 shows the limits on the fitting parameters of this equation that lead to the Chick, Chick–Watson, Hom's, Power Law, and Hom–Power Law inactivation models (Santoro et al., 2007). These models can be used to determine the required time or active concentration required to achieve a given amount of log inactivation under specific water quality conditions. However, PAA decomposition kinetics must be applied to determine the *PAA dose* required to achieve a given active concentration at a given time. It should be noted, however, that the models discussed are simply a mathematical way to fit microbial inactivation data and are specific to a particular set of conditions. As such, there is not a mechanistic basis for the inactivation observed.

TABLE 3.4 Inactivation model comparison (Santoro et al., 2007).

Model parameters	Chick	Chick–Watson	Hom's	Power Law	Hom–Power Law
k′	$k' \neq 0$	$k' \neq 0$	$k' \neq 0$	$k' \neq 0$	$k' \neq 0$
n	0	$n \neq 0$	$n \neq 0$	$n \neq 0$	$n \neq 0$
m	1	1	$m \neq 1$	1	$m \neq 1$
x	1	1	1	$x \neq 1$	$x \neq 1$
dN/dt	$-k'N$	$-k'C^nN$	$-k'C^nmNt^{m-1}$	$-k'C^nN^x$	$-k'C^nmN^xt^{m-1}$

Rossi et al. (2007) compared the fit of several disinfection models, including the Selleck model (1978), the Chick–Watson model (1908), the Hom's model (1972), and the S-model (1998), to experimental data from full-scale water resource recovery facilities (WRRFs). Neither the Selleck model nor the Chick–Watson model fit the data, but the S-model and the Hom's model fit the data within acceptable limits. The S-model fit the data better for PAA doses less than 5 mg/L, whereas the modified Hom's model fit the data better for PAA doses greater than 5 mg/L.

Data from the aforementioned study indicated that bacterial inactivation was more sensitive to residual concentration than contact time, although the latter became more important at lower PAA doses. In this study, microbes damaged by PAA were unable to repair and regrow up to 5 hours after the PAA residual was quenched, suggesting that the damage to microorganisms is permanent and eventually results in cell death.

When PAA is used in secondary effluent, the pseudo-mechanistic double exponential model (Equation 3.15) is especially useful in describing inactivation kinetics (Santoro et al., 2015). In this model, log inactivation is related directly to integral CT. The bacterial populations are split into two distinct subpopulations, each linked to a separate decay coefficient. The subpopulation k_d, which can be considered non-particle associated (i.e., free floating), has a fast inactivation rate, and the other subpopulation, k_p, which can be considered particle-associated, has a slower inactivation rate that is limited by diffusion.

$$N_t = N_0(1 - \beta)e^{-k_d CT_{PAA}{m}} + N_0(\beta)e^{-k_p CT_{PAA}} \tag{3.15}$$

Where

N_t = the microbial concentration at time, t;
N_0 = the microbial concentration at time, 0;
β = the percent of the microbial subpopulation that is particle associated;
k_p = the decay coefficient associated with the particle-associated subpopulation;
k_d = the decay coefficient associated with the free-floating subpopulation;
CT_{PAA} = the integral contact time; and
m = coefficient to describe "shouldering" effect.

A correlation analysis performed on the parameters in the double exponential model and the decomposition model showed that the particle-associated fraction of the bacterial population was directly correlated with kinetic parameters k and D. This indicates that poorer wastewater quality results in a greater population of particle-associated bacteria, a greater instantaneous demand, and faster decomposition kinetics. The opposite is true as wastewater quality improves. For filtered wastewaters (or other situations) in which the microorganism population is more homogenous, this model can be simplified to describe the relationship between log inactivation and contact time using a single inactivation decay coefficient.

None of these models are definitively mechanistic, and the application of inactivation models to experimental data is essentially a curve-fitting exercise; therefore, it is acceptable to use whichever model fits the data best to determine the required PAA dose for a specific application and water quality. However, determining the instantaneous demand and the decay coefficient is essential for correct dosing. In addition, when collecting data, it is important to gather enough information to generate a robust model that is valid over the entire range of operational conditions.

2.4 Disinfection Using a Combination of Ultraviolet and Peracetic Acid

In the last 20 years, numerous studies have been performed investigating potential synergies between UV and PAA disinfection processes. A number of researchers have reported on the benefits of combining of UV and PAA to enhance the disinfection of municipal wastewater (U.S. Patent No. 20150005379, 2015; Caretti & Lubello, 2003; Gonzalez et al., 2012; Koivunen & Heinonen-Tanski, 2005b; Lubello et al., 2002; Rajala-Mustonen, Toivola, & Heinonen-Tanski, 1997; U.S. Patent No. 20100176066, 2010). However, the exact mechanism for this enhancement is not clear, and there is no general consensus on how the combination of UV and PAA can result in an enhanced disinfection process. It has been reported that PAA combined with high UV doses (1 to 2 orders of magnitude greater than UV doses used in wastewater disinfection) increased disinfection as a result of an advanced oxidation process, resulting from the photolysis of the O–O bond in the PAA molecule, generating a hydroxyl radical (·OH) (Caretti & Lubello, 2003; Lubello, Caretti, & Gori, 2002).

Before discussing UV disinfection, it is important to define UV dose. In this text, UV dose is defined as the amount of UV irradiation absorbed by an exposed population of microbes, typically in units of mJ/cm^2 (mJ/cm^2 = 1000 $\mu W/cm^2$ per second). It is often assumed to implicitly represent total absorbed dose, which is more accurately described as fluence. However, *dose* is the term that is more commonly used in the industry and, therefore, is used in this discussion.

While investigating the combination of UV and PAA, Lubello et al. (2002) found that neither a PAA concentration between 2 and 8 mg/L nor a UV dose of 120 to 300 mJ/cm^2 were able to reach the target disinfection levels; however, when a PAA concentration of 2 mg/L was applied immediately before a UV dose of 192 mJ/cm^2, over 4 logs inactivation of total coliforms were achieved. However, Gonzalez et al. (2012) reported that when PAA and UV irradiation were combined, at UV doses of 13 mJ/cm^2 more relevant to wastewater disinfection, there was no synergistic benefit observed of adding PAA before or after UV irradiation. This highlights the ambiguity in the current understanding of the synergistic effects of UV and PAA treatments.

If synergies do occur between UV and PAA disinfection, a few mechanisms can be hypothesized. Where PAA precedes UV, it has been proposed that PAA can destabilize or disrupt particle-associated bacteria, thereby eliminating shielding mechanisms and enhancing performance of downstream UV disinfection. Another hypothesis, which will be critically discussed later in this section, is that equilibrium PAA solutions can be photolyzed when exposed to UV irradiation, resulting in the formation of radical species that can enhance disinfection performance. In summary, while some studies have purported a synergy between UV and PAA, resulting in enhanced disinfection performance, more researchers have merely observed an additive effect.

Even without disinfection synergies, there may be other advantages to combined disinfection. The combined use of UV and PAA may lead to an integrated, environmentally friendly disinfection solution that allows optimization of both capital and operating costs while minimizing environmental impacts. UV systems offer the benefit of a chemical-free solution, typically with the lowest overall life-cycle cost. Recently, Murray et al. (2016) reported on the use of UV and PAA disinfectants, and their combinations, for a very challenging secondary effluent. Murray et al. (2016) performed 20-year net present value (NPV) calculations for various combinations of UV and PAA required to achieve a disinfection target of 126 CFU/100 mL in a low-quality, highly colored secondary effluent. The study found that while UV alone always offered the lowest 20-year NPV, there is still a benefit to combined disinfection if a utility had a capital budget constraint. It was observed that a combined approach could meet disinfection targets while offering a resilient and cost-effective approach that balanced capital and operating costs. Thus, irrespective of whether or not there are mechanistic pathways that allow for a synergistic UV and PAA disinfection process, there are clear operational and financial synergies.

2.4.1 Photochemistry of Peracetic Acid Photolysis

Equilibrium solutions of PAA contain a combination of PAA, H_2O_2, acetic acid, and water. Of these four chemicals, under some conditions PAA and H_2O_2 can be photolyzed by UV irradiation of wastewater at 253.7 nm, leading to the generation of radical species including hydroxyl radicals and carbon-centered radicals.

The first law of photochemistry states that in order for a photochemical reaction to occur, a molecule must first absorb a photon (Bolton & Cotton, 2008). Beer's law relates the attenuation of light to the properties of a material and states that the concentration of a chemical is directly proportional to the absorbance of a solution (Patnaik, 2008).

The photochemical parameter used to quantify the ability of a chemical to absorb light at a specific wavelength is the molar extinction coefficient (ε, L/mol/cm). At the typical pH of wastewater (pH 7 to 8), the molar extinction coefficient of H_2O_2 at 254 nm is 19.6 L/mol/cm (Buxton et al., 1988). Under the same pH range, the molar extinction coefficient of PAA at 254 nm has been reported as 10.0 L/mol/cm (Cai et al., 2017). Given these molar extinction coefficients and the typical equilibrium PAA solutions used in wastewater disinfection, Table 3.5 provides the absorption coefficients of PAA and H_2O_2 in concentrated commercial solutions typically used in wastewater disinfection. Note that when diluted in wastewater, the molar extinction coefficient remains unchanged, but the absorption coefficient will be much lower because of the reduction in molar concentration. With that said, it can be assumed that the relative concentration of PAA to H_2O_2 remains unchanged when diluted. It is then observed that for a 15% PAA solution, H_2O_2 will absorb relatively more photons than PAA; however, for 22% PAA, where the H_2O_2 concentration is relatively low, PAA is relatively the greater absorber. It should be noted that the relative concentrations of PAA and H_2O_2 could be different when added to the wastewater because of different instantaneous reactions with the water matrix as well as residual decays.

After a photon is absorbed, a photochemical reaction may occur. The measure of how many photochemical products are formed per number of photons absorbed is known as the quantum yield. Cai et al. (2017) reported the quantum yield at 254 nm of PAA as 1.2 mol/Einstein at neutral pH.

TABLE 3.5 Absorbance at 254 nm of H_2O_2 and PAA in peracetic acid formulations typically used in wastewater disinfection.

	Molar concentration (mol/L)		Molar extinction coefficient (L/mol/cm)		Absorption coefficient (cm^{-1})	
	PAA	H_2O_2	PAA	H_2O_2	PAA	H_2O_2
12% PAA, 20% H_2O_2	1.58	5.88	10.0	19.6	15.8	115.3
15% PAA, 23% H_2O_2	1.97	6.76	10.0	19.6	19.7	132.6
22% PAA, 5% H_2O_2	2.89	1.47	10.0	19.6	28.9	28.8

It was argued that because this value is over twice the quantum yield at 254 nm of H_2O_2 (0.5 mol/Einstein), photolysis of PAA may generate more hydroxyl radicals than photolysis of H_2O_2. However, it should be noted that as per the chemical reaction expression, each mole of H_2O_2 can yield 2 mol of hydroxyl radicals, and so, the apparent quantum yield of H_2O_2 photolysis is 1.0. In contrast, each mole of PAA can only yield 1 mol of hydroxyl radical. Therefore, from quantum yield, and efficiency, perspective PAA and H_2O_2 are similar in terms of their potential to generate hydroxyl radicals under exposure to 254-nm UV irradiation.

It should be noted that although radicals can form from the photolysis of equilibrium PAA solutions, this is a direct function of the oxidant concentration as well as the UV dose applied. Moreover, because radical species are highly reactive and non-selective, not all generated radicals react with the target species. Rather, the large majority of radicals generated are "scavenged" by background organic and inorganic material in a wastewater matrix (Ahuja, 2014). Using a simple photochemical model and assuming a steady-state hydroxyl radical concentration, one can understand the relevance of the radical-driven disinfection process.

Therefore, given the low potential for radical generation resulting from relatively low UV doses and PAA concentrations applied in wastewater disinfection and the high potential for radical scavenging caused by the high concentration of organics and inorganics in wastewater, it is impractical to consider that combining UV and PAA can be considered as a process for generating radical species for disinfection at doses commonly applied at WRRFs. However, a combined UV and PAA disinfection process does have potential complementary benefits as well as operational and cost synergies.

2.4.2 Complementary Inactivation Kinetics

The selection of order and sizing of UV and PAA disinfection processes is dependent on the disinfection kinetics of each disinfectant alone as well as potential synergies. For example, an experimental methodology to determine order of disinfection and required doses may include consideration of (a) the dose response curve of each individual disinfectant; (b) the correlation of microbial log inactivation to relevant treatment constraint (e.g., cost, time required, energy demand, footprint, etc.); (c) an algorithm to determine the sequence and amount of disinfectant dose, where the disinfectant with the fastest kinetics (i.e., greatest slope) is always selected in order to minimize the treatment constraint; and (d) dose response curves generated for the combinations of UV and peracid disinfectants, with each disinfectant applied sequentially, in vice-versa order, and simultaneously. If synergies are observed for a particular sequence, this sequence is selected.

Basically, by knowing the kinetics of each disinfectant alone and using the concept of multi-target disinfection (i.e., the ability of two or more disinfectants to inactivate the same bacterial population via two or more mechanisms or mode of actions), one can determine the most cost-effective sequence of the two disinfectants, and the relative amount that minimizes overall treatment constraint (e.g., cost, footprint, energy demand, or time).

Figure 3.4 presents several reported examples illustrating the application of this method to maximize disinfection efficiency using UV and PAA. This figure illustrates the dose response curves for UV only, PAA only, and combined disinfection, for both the fast and slow inactivation regimes, in four different disinfection scenarios. In the first scenario, the most efficient method is to apply PAA first and UV second. In the second scenario, the best combination is UV first and PAA second. In the third scenario, the best combination is PAA first and UV second, followed by a second PAA dose. In the last scenario, different populations of microorganisms display different resistances to UV and PAA disinfection. Neither disinfectant alone can reach the treatment goal in a feasible manner. Here, the best combination is to apply UV disinfection until tailing occurs, thereby only leaving behind UV-resistant microorganisms. Then, PAA is applied to inactivate the remaining microorganisms that are susceptible to PAA.

2.4.3 Modelling Dual Disinfection Inactivation Kinetics

Murray et al. (2016) conducted UV+PAA tests where UV doses of 10, 15, and 20 mJ/cm^2 were applied before PAA, which was applied at contact times ranging from 5 to 25 mg·min/L. The experimental results indicated a clear, consistent, and logical trend in that, (a) for the initial UV treatment, increasing UV dose resulted in reduced concentration of viable *E. coli*, and (b) for the secondary PAA treatment, increasing PAA contact times resulted in either reduced or constant concentration of viable *E. coli*.

The authors went on to develop a pseudo-mechanistic model where populations of microbes were separated and modelled based on their susceptibility to a single disinfectant (UV or PAA), with each population having its own inactivation kinetics. For the cases where two disinfectants were applied (UV and PAA), the microbes were modelled as four populations: A0 represented a population easy to inactivate by UV and PAA; B0 represented a population easy to inactivate by UV and hard to inactivate by PAA; C0 represented a population hard to inactivate by UV and easy to inactivate by PAA; and D0 represented a population hard to inactivate by UV and hard to inactivate by PAA. Figure 3.5 provides a conceptual illustration of this mechanistic approach.

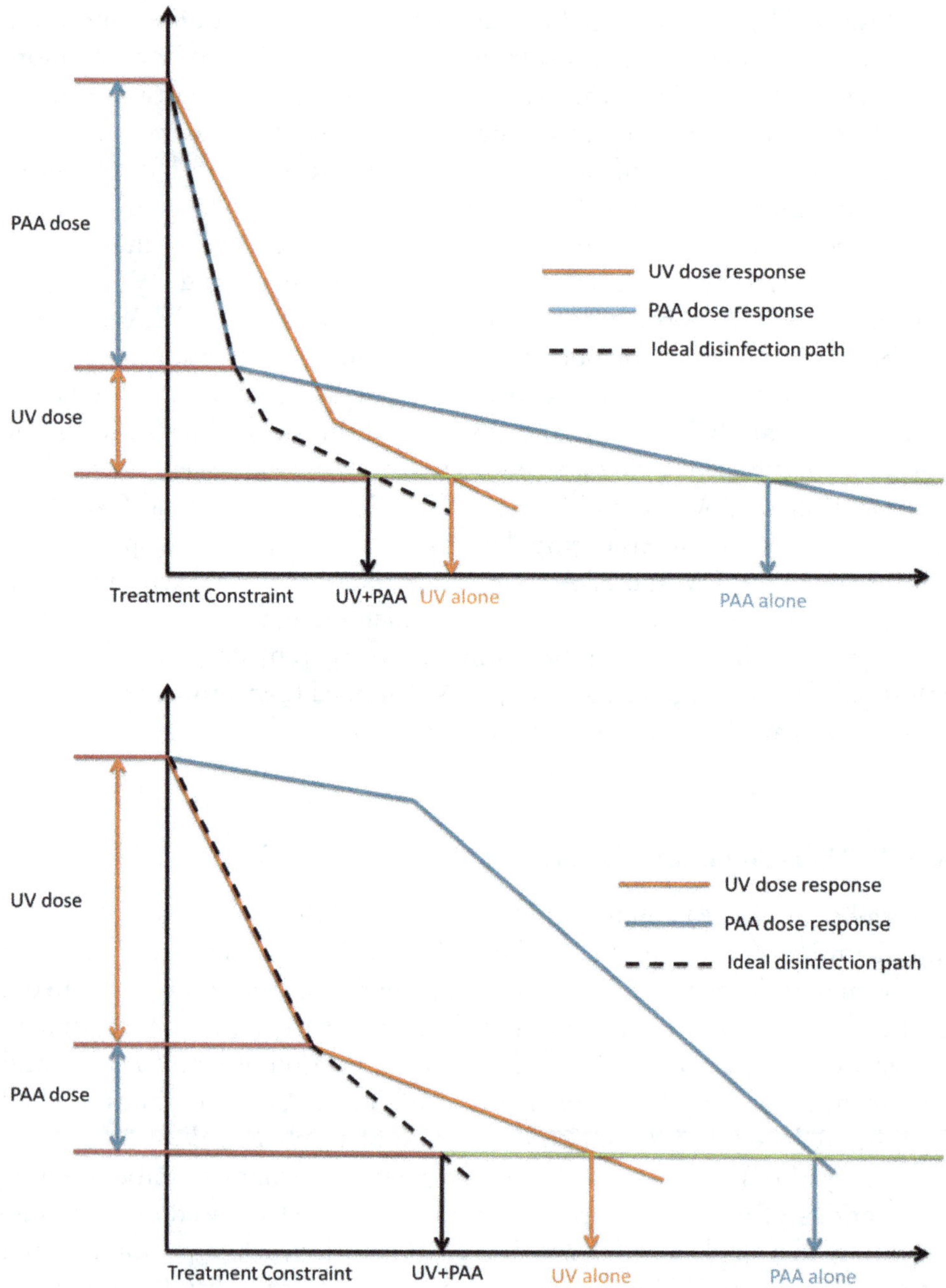

FIGURE 3.4 Four scenarios for sizing and selecting a combined UV and peracid disinfection process (Murray et al., 2016).

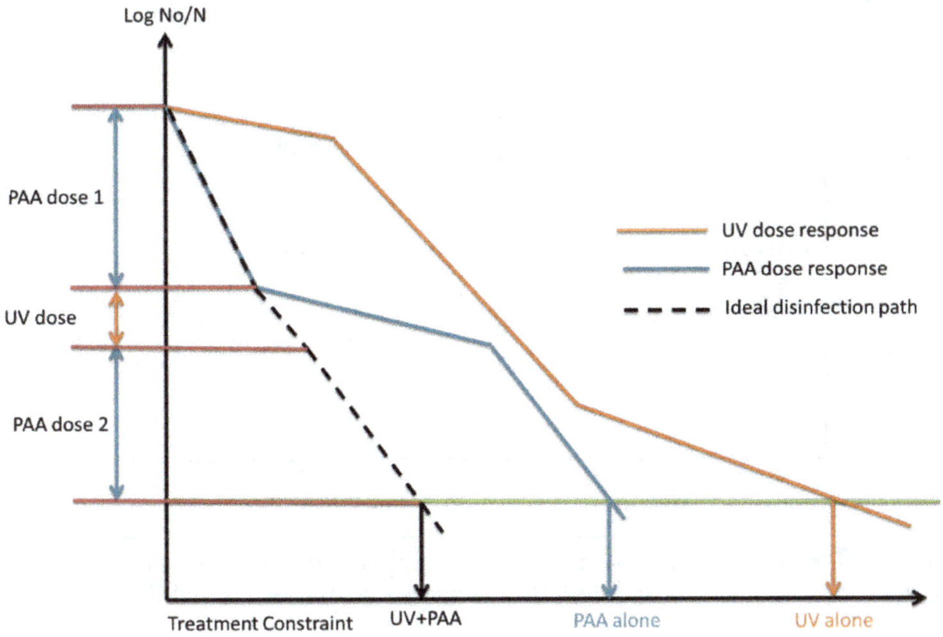

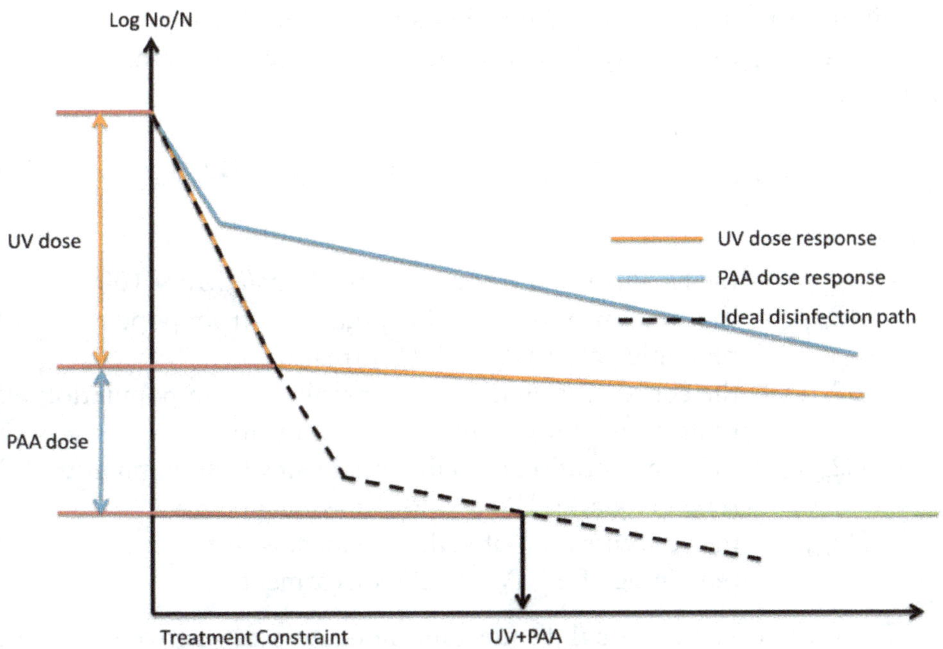

FIGURE 3.4 Four scenarios for sizing and selecting a combined UV and peracid disinfection process (Murray et al., 2016) (*Continued*).

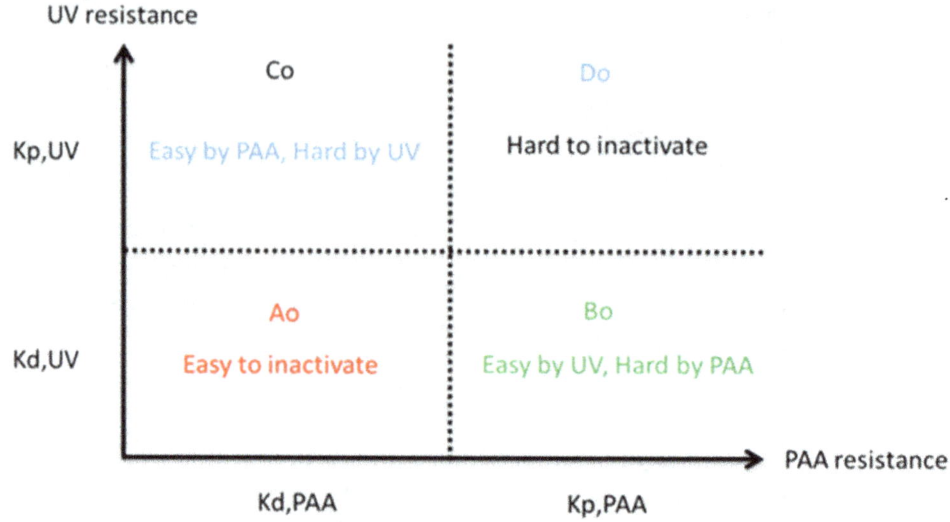

FIGURE 3.5 Four population mechanistic approach to model a dual disinfectant system consisting of UV and PAA disinfectants (Murray et al., 2016).

Figure 3.6 gives formulas for this system. The following equation (3.16) was used to quantify the concentration of culturable organisms after disinfection:

$$N_{total,viable} = A2_{viable} + B2_{viable} + C2_{viable} + D2_{viable} \tag{3.16}$$

Where

$N_{total,viable}$ = the total concentration of viable *E. coli*, MPN/100 mL;
$A2_{viable}$ = the concentration of viable organisms from population A0 remaining after UV and PAA treatment;
$B2_{viable}$ = the concentration of viable organisms from population B0 remaining after UV and PAA treatment;
$C2_{viable}$ = the concentration of viable organisms from population C0 remaining after UV and PAA treatment; and
$D2_{viable}$ = the concentration of viable organisms from population D0 remaining after UV and PAA treatment.

The four population, dual disinfectant model was fitted to the UV-only, PAA-only, and UV+PAA experimental data to estimate the model parameters A0, B0, C0, D0, kd,UV, kd,PAA, kp,UV, and kp,PAA. The resulting model is shown with observed results for UV only and PAA only (Figure 3.7) and for UV→PAA (Figure 3.8).

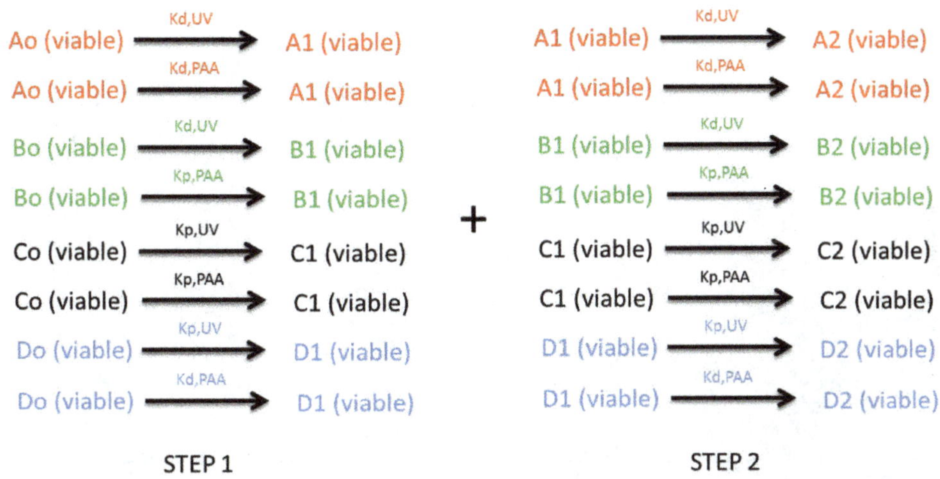

FIGURE 3.6 Formulas representing inactivation products and respective first order rate constants for the proposed four population system with dual disinfectants (Murray et al., 2016).

Figure 3.9 illustrates the comparison between actual results and model-predicted concentrations of culturable *E. coli*. In general, the four populations, dual disinfectant model was able to reasonably predict the effect of UV-only, PAA-only, and UV+PAA disinfection of *E. coli*.

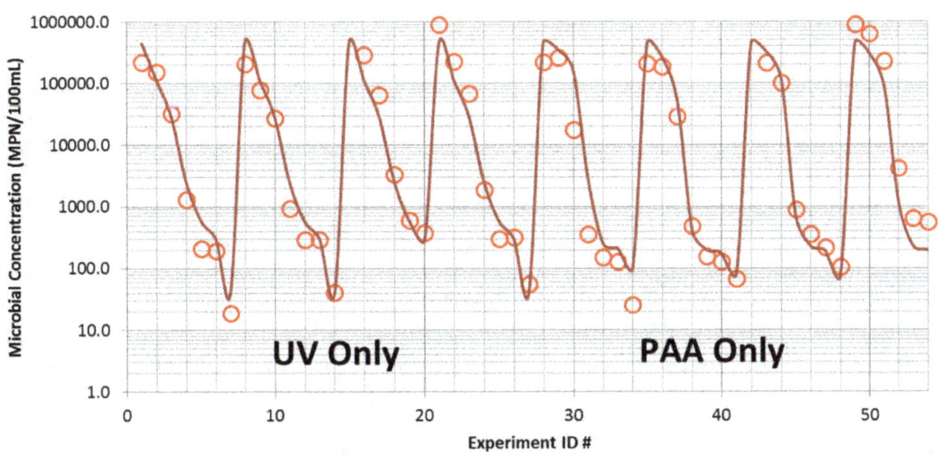

FIGURE 3.7 Experimental results from inactivation using UV or PAA alone (circles) and predicted values using the four population, dual disinfectant model (Murray et al., 2016).

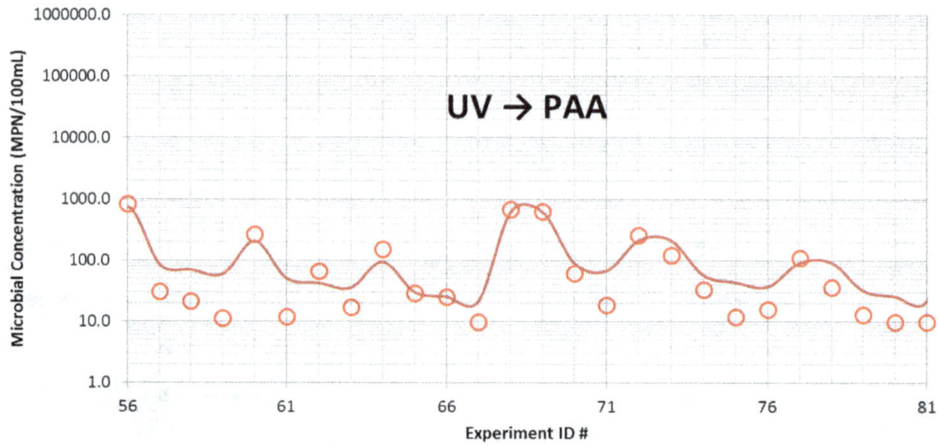

FIGURE 3.8 Experimental results from inactivation using UV→PAA (circles) and predicted values using the four population, dual disinfectant model (Murray et al., 2016).

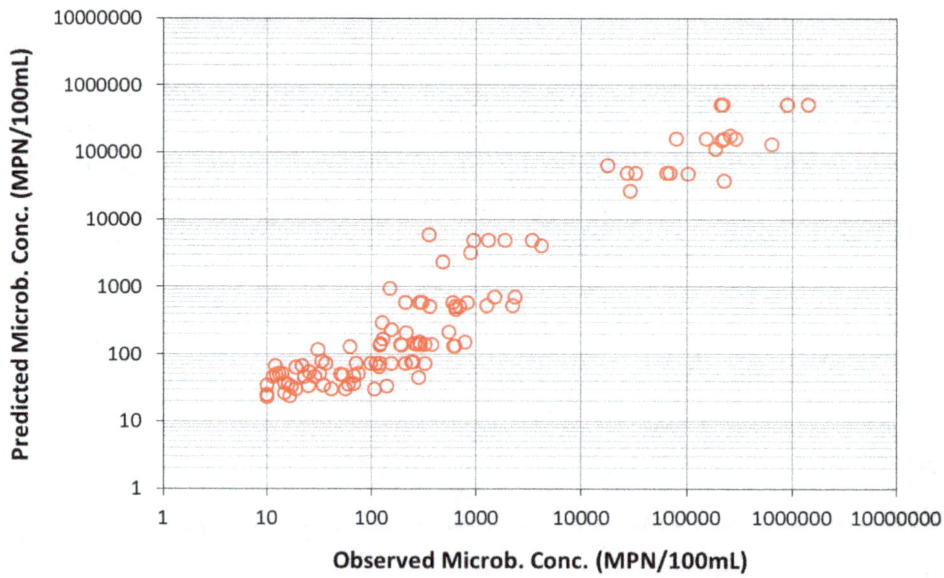

FIGURE 3.9 Comparison of the observed concentrations of viable *E. coli* after disinfection (UV-only, PAA-only, and UV+PAA) and the four population, dual disinfectant model predicted concentration of viable *E. coli* (Murray et al., 2016).

2.4.4 Operational and Economic Benefits of Ultraviolet Plus Peracetic Acid Disinfection

Murray et al. (2016) used the aforementioned model to determine the combination of UV doses and PAA contact time that would achieve a disinfection target of 63 MPN/100 mL. Figure 3.10 illustrates the model-predicted combinations of disinfectants that would be required to meet the disinfection

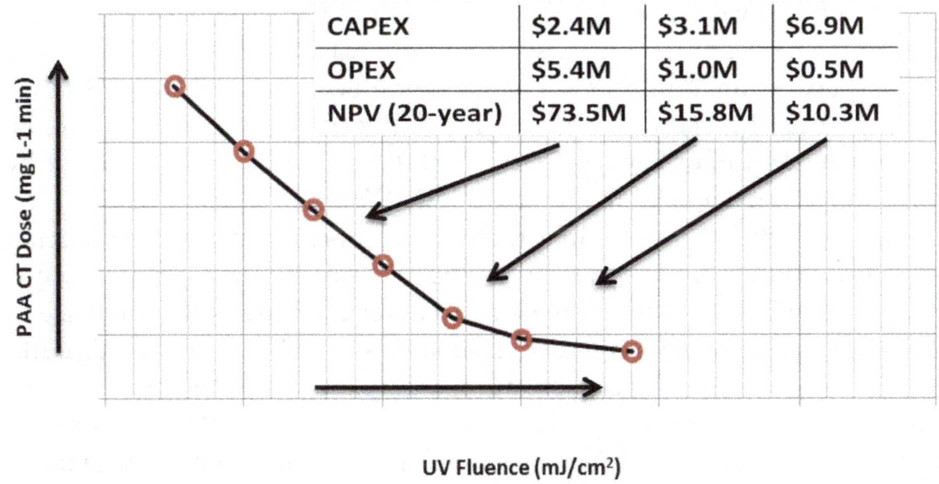

FIGURE 3.10 Model-predicted combinations of PAA dose and UV fluence required to achieve an *E. coli* disinfection target of 63 MPN/100 mL when applying the sequential UV→PAA treatment process (Murray et al., 2016).

target. This plot shows that as the delivered UV dose is decreased, PAA can be brought online to supplement UV and still meet the disinfection target. Potential situations where the delivered UV dose could drop, and PAA could be brought online, include (a) storm events leading to flows above the design capacity, (b) a drop in ultraviolet transmittance (UVT) as a result of an industrial discharge, and/or (c) secondary clarifier upset conditions leading to an excess escape of suspended solids. The trendline in Figure 3.10 shows the predicted PAA contact time (CT) for varying pre-UV doses. This correlation could be used to evaluate the economics, including operating and capital costs, associated with various UV and PAA combinations.

2.5 Water Quality Impacts on Peracetic Acid

The decomposition rate of PAA as well as its disinfection efficacy are strongly affected by water quality. Some influencing factors are pH and water matrix composition, including the presence of transition metals, organic matter, suspended solids, and water hardness (Domínguez Henao, Cascio, Turolla, & Antonelli, 2018; Domínguez Henao, Delli Compagni, Turolla, & Antonelli, 2018). The following subsections discuss the main water characteristics impacting PAA disinfection of wastewater in detail.

2.5.1 pH

Various authors have reported the effect of pH on PAA decomposition rate and disinfection efficacy (Kitis, 2004; Luukkonen et al., 2014; Luukkonen & Pehkonen, 2017; Yuan, Ni, & Van Heinegen, 1997a, 1997b). As detailed in Chapter 2, since the pK_a of PAA is 8.2, at pH higher than this value the equilibrium moves toward its dissociated form (CH_3COOO^-), affecting both the PAA decay rate and bacterial inactivation performance. In a pH range of 5.5 to 8.2, PAA decay is mainly caused by spontaneous decomposition, whereas at pH values between 8.2 to 9.0, PAA decay is mainly driven by spontaneous decomposition and hydrolysis (Yuan et al., 1997a).

However, the PAA decomposition rate is approximately stable in a pH range between 6.5 and 8.2 (Pedersen, Meinelt, & Straus, 2013), which is the typical pH interval of wastewater. Chapter 2 discusses the influence of pH on PAA decomposition in detail. In addition, pH also affects PAA efficacy in terms of microbial inactivation, being more effective under acidic conditions and decreasing at pH values above 9 (Kitis, 2004; Luukkonen et al., 2014). Again, this might be related to PAA dissociation at alkaline pH values because the non-dissociated form is thought to be a more active disinfectant than the dissociated form (Yuan et al., 1997b).

2.5.2 Inorganic Compounds

The presence of some inorganic compounds in water can promote PAA decomposition, depending on the nature of these molecules and, consequently, can determine a reduction in disinfection efficacy. Various authors have addressed efforts to elucidate the role of different inorganic compounds according to their typology, such as transition metals, phosphate, nitrogen compounds, hardness, and salinity. The following subsections describe the role of these compounds on PAA decomposition and disinfection efficacy.

2.5.2.1 Transition Metals

Various studies (Kitis, 2004; Luukkonen et al., 2014; Luukkonen & Pehkonen, 2017; Pedersen et al., 2013; Wagner et al., 2002; Yuan et al., 1997a, 1997b; Zhao et al., 2008) have documented the effect of the presence of transition metals in water on PAA decomposition. According to the study on the decomposition kinetics of PAA carried out by Yuan et al. (1997a, 1997b), reduced metal ions, such as iron (Fe^{2+}), manganese (Mn^{2+}), copper (Cu^{2+}), and cobalt (Co^{2+}), catalyze PAA decomposition via complex redox reactions. These reactions were further assessed in other works by Popov et al. (2005) and Rothbart, Ember, & van Eldik (2012), who reported that a number of different oxidation states of metal ions are involved. Moreover, it has been observed that chelating agents, such as diethylenetriaminepentaacetic acid (DTPA) and diethylenetriaminepentamethylenephosphonic acid (DTMPA), counteract the catalytic effect of these reduced metal ions.

In agreement with these findings, Domínguez Henao, Delli Compagni, Turolla, and Antonelli (2018) observed that the concentration of Fe^{2+} has a significant effect on PAA decomposition under the investigated operating conditions (0.1 to 5 mg/L of Fe^{2+}). This effect can be counteracted by the presence of phosphates that can, indeed, act as chelating agents toward transition metals in water, inhibiting their catalyzing action on PAA decomposition (Aklil, Mouflih, & Sebti, 2004; Domínguez Henao, Delli Compagni, Turolla, & Antonelli, 2018; Elouear et al., 2008; Irani & Morgenthaler, 1963; Klueh & Robinson, 1988; Larson, 1957; Varner et al., 1996).

Despite their catalytic action on PAA decomposition, the presence of transition metals, such as copper (Cu^{2+}) and silver (Ag^{+}), can also enhance the formation of radicals, improving PAA efficiency in terms of microbial inactivation. Orta de Velásquez et al. (2017) observed a synergy between Ag^{+} and PAA in the removal of *V. cholerae, Salmonella* sp., *Shigella, P. aeruginosa*, helminth eggs, and fecal coliforms. Luna-Pabello et al. (2009) reported similar synergistic effects for systems combining Ag^{+}/PAA and Ag^{+}/Cu^{2+}/PAA for the removal of helminth eggs. The authors highlighted that one major advantage of the addition of Cu^{2+} and/or Ag^{+} ions is a decrease in the required contact time (Luna-Pabello et al., 2009; Luukkonen & Pehkonen, 2017; Orta de Velásquez et al., 2017). However, the addition of metal ions can cause secondary contamination, impairing water quality. Furthermore, the economic feasibility of the use of Cu^{2+} and Ag^{+} should be evaluated (Luukkonen & Pehkonen, 2017).

2.5.2.2 Phosphates

An interesting feature of phosphates, besides their role as chelating agents toward transition metals in solution, is their potential capacity to inhibit inactivation, as observed by Mattle et al. (2011) on bacteriophage MS2. In detail, the authors reported that high phosphate concentrations (400 mM; approximately 746 mg/L) reduce the inactivation extent by a factor of 5.1 compared to lower phosphate concentrations (15 mM; approximately 28 mg/L). The protective effect of phosphate was attributed to its capacity to quench free radicals (Booth & Lester, 1995; Mattle et al., 2011).

2.5.2.3 Nitrogen Compounds

Despite the ubiquitous presence of inorganic nitrogen-containing compounds in different types of water, such as biologically treated effluents, little work has addressed their effect on PAA disinfection. Domínguez Henao, Delli Compagni, Turolla, and Antonelli (2018) investigated the role of ammonia nitrogen (NH_4^{+}), nitrate (NO_3^{-}), and nitrite (NO_2^{-}) on PAA decay. These authors observed a minor effect of these compounds on PAA oxidative demand; however, this effect was not statistically significant under the investigated operating conditions (0.5–25 mg/L as N for NH_4^+ and NO_3^- and 0.1–1.0 mg/L as N for NO_2^-).

2.5.2.4 Hardness and Salinity

A faster decomposition of PAA in water with high salinity (ionic strength) has been suggested in a number of studies (European Centre for Ecotoxicology and Toxicology of Chemicals, 2001; Luukkonen et al., 2014; Soares Cavallini, De Campos, Beber De Souza, & De Sousa Vidal, 2013). Liu et al. (2014) investigated the effect of salinity and water hardness on PAA decomposition in detail by using sodium chloride (NaCl) and sea salt to reach concentrations of 1% (10 g/L) and 3% (30 g/L) in distilled water. The authors found that salinity has a relevant effect on PAA and, interestingly, sea salt led to higher PAA degradation rates than pure NaCl. Indeed, the main components of sea salt used were NaCl and potassium chloride (KCl), which represent around 85% w/w of its composition, whereas the remaining 15% w/w is composed of SO_4^{2-}, Mg^{2+}, and Ca^{2+}. This suggests that the ionic composition also plays a relevant role on PAA decomposition. The authors attribute this effect to Mg^{2+}/Ca^{2+} and Na^+/K^+ ratios because higher Mg^{2+}/Ca^{2+}ratios (85% in standard dilution water) led to slower kinetics, whereas higher Na^+/K^+ ratios (85% in sea salt) resulted in faster reactions. Regarding water hardness, the same authors found that water hardness (4 to 460 mg/L Ca^{2+} and Mg^{2+}) had only a slight impact on PAA decomposition. No other research on PAA mentions hardness as one of the main parameters driving its decomposition in water.

2.5.3 Organic Matter

Several authors (Domínguez Henao, Delli Compagni, Turolla, & Antonelli, 2018; Gehr & Cochrane, 2002; Koivunen & Heinonen-Tanski, 2005b; Liu et al., 2014; Pedersen et al., 2013) have highlighted that the presence of high loads of organic matter in water leads to a significant PAA consumption. For instance, municipal wastewater after primary treatment characterized by high chemical oxygen demand (COD) concentrations (124 to 240 mg/L), TSS (16 to 45 mg/L), and turbidity (16 to 31 NTU) exerts a significant PAA oxidative demand (Gehr & Cochrane, 2002).

Moreover, PAA residuals are typically very low or absent after the disinfection of primary effluents with high COD, TSS, and turbidity levels (Koivunen & Heinonen-Tanski, 2005a), whereas they are typically present after the disinfection of secondary and tertiary effluents. Pedersen et al. (2013) studied the effect of different levels of organic matter content as COD concentration (up to 70.8 mg/L of COD) on PAA decomposition. The authors observed a significant and positive correlation between COD concentration and PAA oxidative demand. They also observed the effect of organics on PAA oxidative demand, adopting dissolved organic carbon (DOC) as an indicator of organic matter. In agreement with previously cited works, Liu et al. (2014) observed an increasing trend of PAA oxidative demand at increasing DOC concentrations in a range between 8 and 24 mg/L, using a synthetic media prepared from commercial humic acids.

Organic matter is a heterogeneous and complex mixture; thus, most research work has adopted macroparameters, such as COD and DOC, as indicators of organic content. However, different organic macromolecules, namely carbohydrates, lipids, and proteins, are expected to exert different effects on PAA oxidative demand because of their different nature and composition. Domínguez Henao, Delli Compagni, Turolla, & Antonelli (2018) studied the effect of the different groups of organic macromolecules on PAA consumption, evaluating them by their contribution as COD (10 to 35 mg/L). Different types of macromolecules displayed different effects on PAA oxidative demand: Proteins were the only macromolecules that exerted a significant PAA demand, whereas carbohydrates and lipids had only a minor effect, which provided evidence that not all components of COD result in significant PAA consumption.

Thus, the authors hypothesized that PAA has a specific oxidative action on proteins, which can entail PAA consumption. Indeed, previous works have also addressed the specific oxidative capacity of PAA toward amino acids. For instance, Kerkaert et al. (2011) observed a rapid oxidation of cysteine, tryptophan, and methionine present in dairy proteins. In agreement with this finding, Finnegan et al. (2010) observed that PAA specifically oxidizes amino acids in the cell wall of microorganisms, degrading, for instance, bovine serum albumin and enzymes.

2.5.4 Suspended Solids and Particulate Matter

Suspended solids and particulate matter in wastewater have a heterogeneous composition, which varies with the characteristics of the effluent, depending mainly on the source (domestic, industrial, agricultural, stormwater); thus, a detailed characterization of their composition is very challenging (Domínguez Henao, Cascio, Turolla, & Antonelli, 2018). However, regardless of the relevant differences in composition, several works highlighted the detrimental effect of TSS on the decomposition and disinfection efficacy of PAA (Chhetri et al., 2014; Chhetri, Bonnerup, & Andersen, 2016; Domínguez Henao, Cascio, Turolla, & Antonelli, 2018; Falsanisi et al., 2008; Koivunen & Heinonen-Tanski, 2005a; Lazarova et al., 1998; Lefevre et al.,

1992; McFadden et al., 2017; Sánchez-Ruiz, Martinez-Royano, & Tejero-Monzon, 1995; Stampi, De Luca, & Zanetti, 2001).

In addition, a relevant feature of TSS regarding PAA disinfection is related to their structure. As particles, TSS are not smooth and rigid, but rather irregular, sponge-like matrixes with pores in which bacteria can be shielded (Dietrich et al., 2003). Accordingly, TSS can impact PAA disinfection via two mechanisms, namely determining PAA's oxidative demand (and associated reduction of PAA's actual disinfecting dose) and acting as protective shields for bacteria against the disinfectant (Domínguez Henao, Cascio, Turolla, & Antonelli, 2018).

The detrimental effect of TSS on PAA decomposition has been discussed in various works on PAA disinfection, including Chhetri et al. (2016, 2014), Falsanisi et al. (2008), Koivunen and Heinonen-Tanski (2005a), Lazarova et al. (1998), Lefevre et al. (1992), McFadden et al. (2017), Sánchez-Ruiz et al. (1995), Stampi et al. (2001), and Domínguez Henao, Delli Compagni, Turolla, and Antonelli (2018). Water matrixes with a high content of TSS, such as combined sewer overflows (CSOs) and primary effluents, require higher PAA concentrations for disinfection than secondary and tertiary effluents (Chhetri et al., 2016; Gehr & Cochrane, 2002; Koivunen & Heinonen-Tanski, 2005a; Luukkonen et al., 2014). Moreover, it has been observed that the pretreatment of CSOs and the removal of suspended solids decrease the required disinfectant dose (Chhetri et al., 2016). The effect of different concentrations of TSS (i.e., 5, 40, 80, 120, and 160 mg/L) on PAA decay was studied by Domínguez Henao, Delli Compagni, Turolla, and Antonelli (2018), who also evaluated the effect of the soluble matter associated to the TSS. Increasing TSS concentrations leads to higher PAA decomposition rates, which increase significantly when more than 40 mg/L of TSS is present. In addition, the effect of TSS on PAA decomposition was much more relevant than the effect of the associated soluble fraction, and the two components were found to be independent and additive.

Although the presence of TSS and particulate matter results in high decomposition rates of PAA, the most relevant effect of TSS on PAA disinfection is often related to the inactivation of bacteria. Indeed, it has been largely documented that TSS impair the disinfection of water disinfection through chemical (chlorine-based compounds, ozone, PAA) or physical methods (UV irradiation) (Tchobanoglous et al., 2014). Because of their morphology, suspended solids and particulate matter can afford protection to microorganisms within their pores (Chhetri et al., 2016), allowing them to resist even high disinfectant dosages (Dietrich, Loge, Ginn, & Başağaoğlu, 2007; Falsanisi et al., 2008).

The effect of the particle size of the solids on this aspect has gained increasing attention over the last decade. Previous works elucidating this aspect (Falsanisi et al., 2008; McFadden et al., 2017) reported that TSS with a particle size between 10 and 120 μm have a minor effect on PAA disinfection, affording a protection to microorganisms of 0.6 logs, whereas particles larger than 120 μm can afford around 1.3 logs of protection. In addition, because bacteria tend to aggregate in natural environments as protection mechanism against extreme conditions, particulate matter can act as a condensation nucleus for bacteria aggregation, resulting in the survival of the microorganisms located in the core of the aggregate (Bohrerova & Linden, 2006; Mir, Morató, & Ribas, 1997; Turolla et al., 2017). This results in the survival of the microorganisms located in the core of the aggregate (Domínguez Henao, Cascio, Turolla, & Antonelli, 2018). Thus, either embedded or aggregated, microorganisms associated to TSS have increased resistance to inactivation by different disinfectants, including PAA, compared to free-swimming microorganisms (Bohrerova & Linden, 2006; Dietrich et al., 2007; LeChevallier, Hassenauer, Camper, & McFeters 1984; Winward, Avery, Stephenson, & Jefferson, 2008).

Lefevre et al. (1992) reported that high TSS concentrations ranging between 10 and 50 mg/L had no significant effect on the disinfection performance of PAA, whereas TSS concentrations over 100 mg/L reduce the extent of inactivation by 1 log in secondary effluents. Other authors (Lazarova et al., 1998; Stampi et al., 2001) observed that the effect on PAA disinfection performance is moderate and constant for TSS concentrations between 10 and 40 mg/L.

3.0 DOSE CONTROL FOR DISINFECTION OF WASTEWATER

3.1 Disinfection Control Strategies

The main aspects of ensuring effective disinfection are via the integration of (a) microbial inactivation/formation kinetics of disinfection byproducts, (b) disinfectant demand (D) and decay (k), and (c) reactor (chemical contact basin [CCB]) hydraulics to an integrated disinfection design framework (Bellamy, Finch, & Haas, 1998). As discussed in previous sections, understanding the water quality and treatment requirements in terms of integrated CT (ICT) dose (i.e., the area under the PAA decay curve) (Santoro et al., 2015) is important for the development of advanced control strategies to maintain constant disinfection performance. Underdosing a disinfectant will result in poor disinfection performance. Conversely, overdosing results in increased operational cost resulting from increased consumption of chemical disinfectant. Moreover, if the disinfecting agent is applied in excess, the formation of undesired disinfection byproducts is enhanced, and the treated water may adversely affect the environment and public health (Richardson, Simmons, & Rice, 2002). Thus, it is important to develop control strategies able to maintain constant ICT dose (ICT-dose pacing), and hence constant disinfection performance.

The most commonly used chemical disinfection control strategies target a constant concentration of the chemical disinfectant using feed-forward flow-pacing control (Kobylinski, Hunter, & Shaw, 2006). The flow-pacing control strategy ignores both contact time and decay (k), hence it does not maintain a constant ICT dose. As a result, the disinfection performance varies with water quantity and quality. Figure 3.11 shows an illustration of the effect of water quantity (flow or contact time) and quality (k) on two control strategies (i.e., flow pacing and ICT-dose pacing).

Flow-pacing control maintains a constant inlet disinfectant concentration (C1) at varied flow rates (Figure 3.11a). Because shorter contact times occur at higher flow rates, maintaining a constant C1 results in a varied ICT dose (i.e., ICT dose will decrease at shorter contact times), as shown by Figure 3.11a and 3.11b. The demand and the decay rate of the disinfectant is ignored by flow-pacing control; hence, the ICT dose varies as function of changes in water quality (Figure 3.11c and 3.11d). In the case of ICT-dose pacing, C1 varies with changes in flow rate to maintain a constant ICT dose (Figure 3.11a and 3.11b). Importantly, ICT-dose control strategy considers water quality

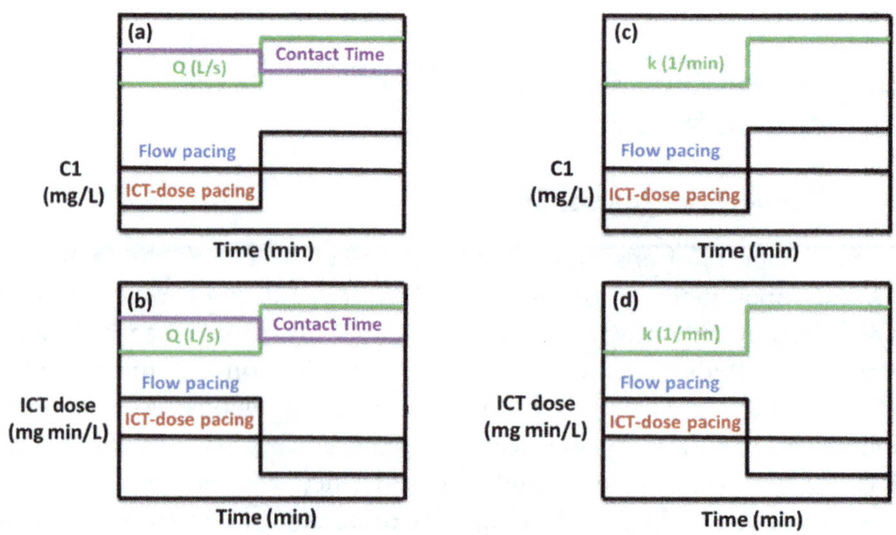

FIGURE 3.11 Comparison of flow-pacing and ICT-dose pacing control strategies. Illustration of effects of flow on (a) disinfectant inlet concentration (C1), and (b) ICT dose. Illustration of effects of disinfectant decay (k) on (c) C1 and (d) ICT dose (Sarathy et al., 2017).

variation by adjusting C1 to account for changes in demand and decay with the goal of maintaining a constant ICT dose (Figure 3.11c and 3.11d).

In wastewater disinfection, the issue of advanced ICT-dose control is the most obvious because the quantity and quality of the water can vary significantly over time based on several factors such as diurnal fluctuations in human activity, industrial discharges, and wet weather events. Under such circumstances, conventional control methodologies such as flow pacing would be inadequate because they are unable to respond appropriately to maintain constant disinfection during these events. A flow-pacing strategy that aims to consider varying water quality using COD and color measurements of the inlet wastewater to feed-forward control the concentration of PAA has been proposed (Block, Morgan, Bell, & Stewart, 2015). However, the aforementioned control strategy still targets a constant inlet PAA concentration. Recently, an innovative ICT-dose pacing control strategy has been developed and compared to flow-pacing control under similar conditions (Sarathy et al., 2017). Section 3.2 presents experimental results of this recently developed ICT-dose strategy.

3.2 Demonstration of Advanced Dose Control Strategy for Peracetic Acid Disinfection

The demonstration of the two control strategies was carried out in a 3-m^3 chemical contact basin (CCB) consisting of four passes. The CCB was located in a WRRF, enabling the use of real-time secondary effluent wastewater as the influent of the CCB. A controller programmed with both conventional flow-pacing logic and the innovative ICT-dose pacing logic was used to control the PAA dosing pump. Peracetic acid was dosed upstream of the pilot-scale serpentine CCB, and PAA probes were installed at the inlet and outlet of the CCB to incorporate wastewater quality changes (i.e., D and k) to the ICT-dose pacing control strategy. The flow of secondary effluent wastewater was adjusted accordingly to simulate relevant wastewater flow variation during a day. Inlet and outlet viable fecal coliforms were measured to monitor disinfection performance.

Initially, a flow-pacing experiment was carried out. The flow ranged from 1.0 L/s (overnight) to 3.5 L/s (morning and evening peak flows), corresponding to hydraulic retention times (HRTs) from 14 to 51 minutes. The ICT dose varies during the day, resulting in high ICT dose at low flows (i.e., longer contact time, higher k) and low ICT dose at high flows (i.e., shorter contact time, lower k) (Figure 3.12). This variation in ICT dose is confirmed through the observation that the concentration of viable fecal coliforms at the outlet is not consistent. That is, the fecal coliform counts increased with increasing flow because the ICT dose was inadequate.

The reduction in the ICT dose is an immediate indication of a potential disinfection permit violation (Figure 3.12). The varied ICT dose, and hence the varied disinfection performnance during flow-pacing operation, was expected because the water quality (D and k) and flow (contact time) are ignored. To ensure disinfection compliance at all times by flow-pacing control, the inlet PAA concentration setpoint has to be set as high as needed to ensure sufficient ICT dose is provided at peak flow/poorest water quality. However, this will result in overdosing of PAA at all lower flows/better water qualities, increasing the operational cost of the process and the formation potential of disinfection byproducts.

In contrast, ICT-dose pacing was able to achieve a constant ICT dose (16 mg·min/L defined by batch experiments) through the numerous flow changes, thereby ensuring consistent disinfection performance (Figure 3.13). Results clearly show that the disinfection performance is driven by the ICT dose (i.e., constant disinfection performance is achieved by constant ICT dose). The small fluctuation of the outlet counts of fecal coliform is caused by the variation of the fecal coliform counts at the inlet (Figure 3.13).

More pilot-scale tests were carried out under flow-pacing operation. The collected data were processed to calculate the amount of PAA needed to treat 1000 m^3 of secondary effluent wastewater. Results showed (Figure 3.14) that the flow-pacing operation required twice the amount of PAA than ICT-dose pacing to maintain a minimum ICT dose of 16 mg·min/L and to ensure disinfection limits were consistently met.

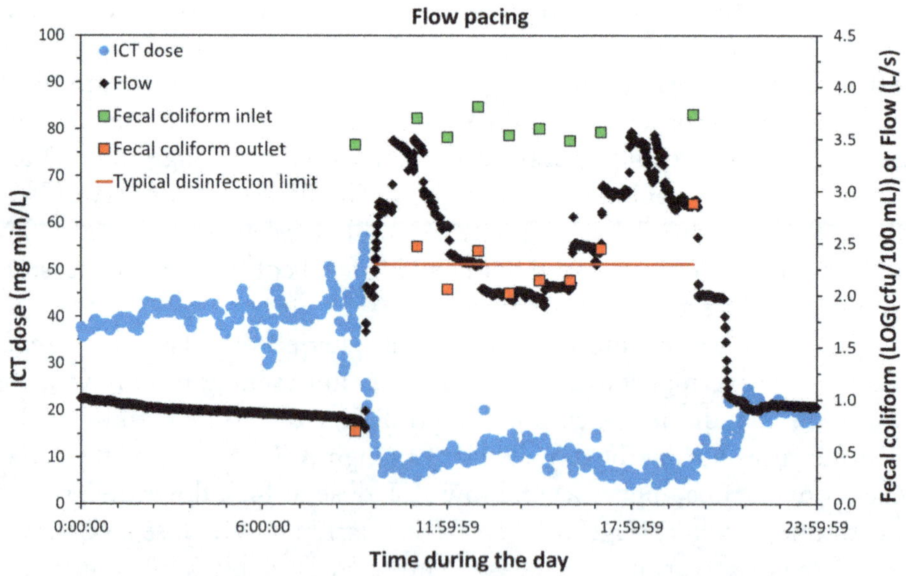

FIGURE 3.12 Performance of flow-pacing control strategy in pilot-scale test (Sarathy et al., 2017).

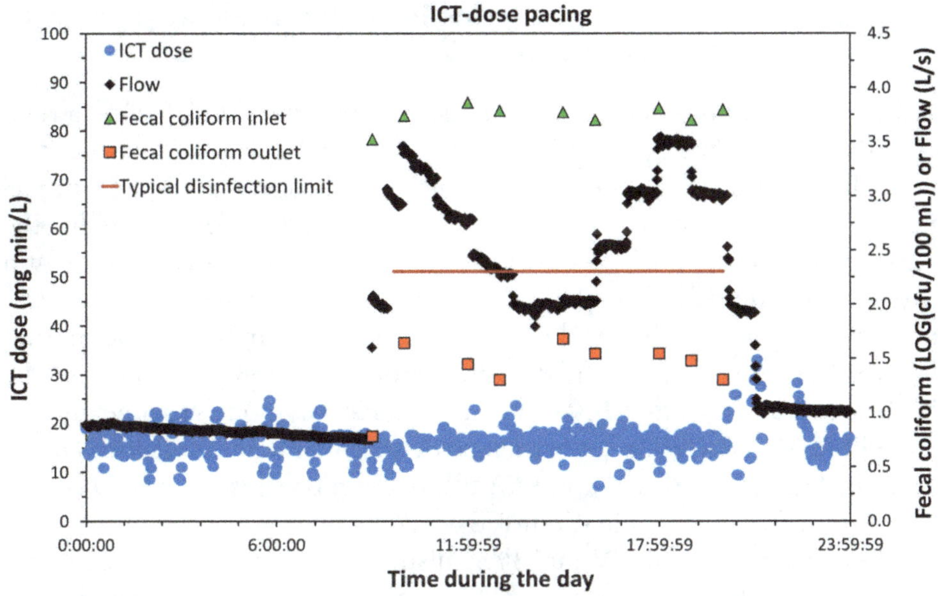

FIGURE 3.13 Performance of ICT-dose pacing control strategy in pilot-scale test (Sarathy et al., 2017).

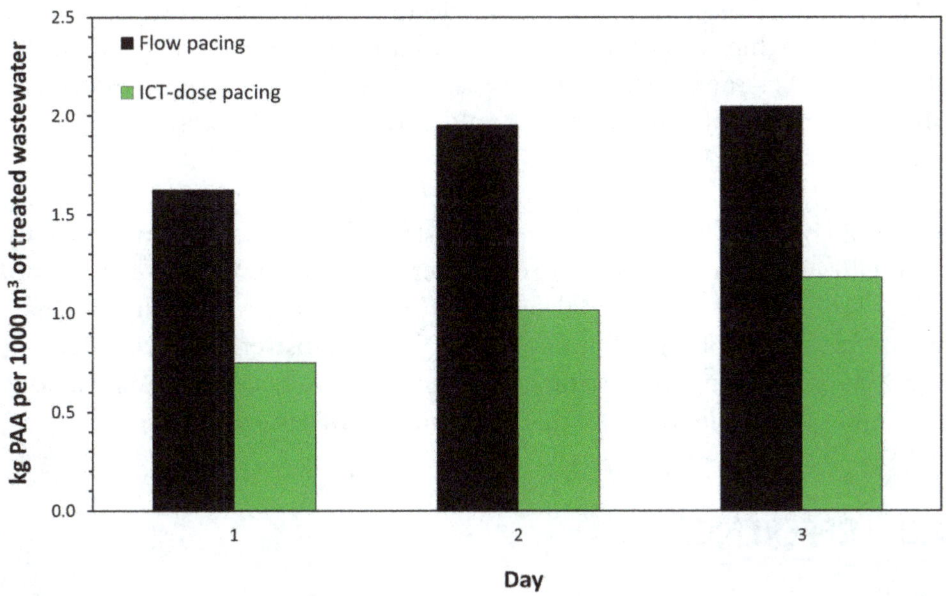

FIGURE 3.14 Amount of PAA required per 1000 m^3 of treated secondary effluent wastewater for each day of pilot-scale tests (Sarathy et al., 2017).

3.3 Dose Control Strategy for Combined Ultraviolet and Peracetic Acid Disinfection Process

Because the PAA disinfection performance is a function of ICT dose, a relationship between the log reduction of specific microbes and ICT dose can be established, using any ICT-dose-based microbial inactivation kinetic model (e.g., double-exponential model). This relationship allows one to predict the concentration of the microbes at the outlet of the disinfection reactor, at a particular ICT dose applied, given the initial concentration of microbes (i.e., concentration of microbes at the inlet of the reactor). It is well known that UV disinfection is also designed and controlled by dose—in this case the UV dose (mJ/cm^2) (Bolton & Linden, 2003). Hence, models can be applied to predict UV disinfection performance of specific microbes at a given UV dose. The fact that both UV and PAA disinfection processes depend on dose is important for the combination of the two technologies.

The combination of UV and PAA disinfection processes in series has been investigated and was previously discussed in Section 2.4 of this chapter. However, a dual-disinfection control strategy is also required for the efficient application of a combined UV and PAA disinfection process. Because dose is the key parameter to control both disinfection processes, an advanced dose control strategy would be able to control the UV dose as well as the ICT dose. This would enable the selection of disinfectant-specific dose setpoints to achieve optimal disinfection pathways considering the nature of the microbes (i.e., their relative resistance to UV and PAA disinfection). Moreover, an intelligent control would transfer disinfection credits from one disinfection process to the other.

Overall, PAA disinfection performance is governed by ICT dose. In other words, a constant ICT dose ensures constant disinfection performance by PAA. Ultraviolet dose is the key parameter to control UV disinfection, too. This enables the use of a dual-disinfection network. An intelligent dual-disinfectant control algorithm could apply the most efficient disinfection setpoints considering the resistance of specific microbes, water quality, flow, UV lamp output, sleeve fouling, and chemical concentration.

4.0 REFERENCES

Ahuja, S. (2014). *Water reclamation and sustainability* (1st ed.). Doi: https://doi.org/10.1016/C2012-0-07359-0

Aklil, A., Mouflih, M., & Sebti, S. (2004). Removal of heavy metal ions from water by using calcined phosphate as a new adsorbent. *Journal of Hazardous Materials*, *112*(3), 183–190.

Antonelli, M., Turolla, A., Nurizzo, C., & Mezzanotte, V. (2013). Peracetic acid for secondary effluent disinfection: A comprehensive performance assessment. *Water Science & Technology, 68*(12), 2638–2644.

Azzellino, A., Antonelli, M., Canziani, R., Malpei, F., Marinetti, M., & Nurizzo, C. (2011). Multivariate modelling of disinfection kinetics: A comparison among three different disinfectants. *Desalination and Water Treatment*, *29*(1–3), 128–139.

Baldry, M. G. C., & Fraser, J. A. L. (1988). Disinfection with peroxides. In K. R. Payner (Ed.), *Industrial biocides* (pp. 91–116). New York, NY: Wiley.

Baldry, M. G. C., & French, M. S. (1989). Activity of peracetic acid against sewage indicator organisms. *Water Science & Technology, 21*(12), 1747–1749.

Bellamy, W. D., Finch, G. R., & Haas, C. N. (1998). *Integrated disinfection design framework*. Denver, CO: American Water Works Association Research Foundation and American Water Works Association.

Bianchini, R., Calucci, L., Caretti, C., Lubello, C., Pinzino, C., & Piscicelli, M. (2002). An EPR study on wastewater disinfection by peracetic acid, hydrogen peroxide and UV radiation. *Annalidi Chimica*, *92*(9), 783–793.

Block, S. S. (2001). Peroxygen compounds. In *Disinfection, sterilization, and preservation* (5th ed.). Philadelphia, PA: Lippincott Williams & Wilkins.

Block, P., Morgan, S., Bell, K., & Stewart, S. (2015). Control strategies for PAA wastewater disinfection at WWTPs with variable effluent quality. *Proceedings of the 88th Annual Water Environment Federation Technical Exhibition and Conference*; Chicago, Illinois, Sept 26–30; Alexandria, VA: Water Environment Federation.

Block, P., & Tran, M. (2015). *U.S. Patent No. 20150005379. Wastewater treatment method.* Washington, DC: U.S. Patent and Trademark Office.

Bohrerova, Z., & Linden, K. G. (2006). Ultraviolet and chlorine disinfection of mycobacterium in wastewater: Effect of aggregation. *Water Environmental Research*, *78*(6), 565–571.

Bolton, J. R., & Cotton, C. A. (2008). *The ultraviolet disinfection handbook*. Denver, CO: American Water Works Association.

Bolton, J. R., & Linden, K. G. (2003). Standardization of methods for fluence (UV dose) determination in bench-scale UV experiments. *Journal of Environmental Engineering,* 129, 209–215.

Booth, R. A., & Lester, J. N. (1995). The potential formation of halogenated by-products during peracetic acid treatment of final sewage effluent. *Water Research*, *29*(7), 1793–1801.

Budde, F. E., & Vineyard, M. K. (2010). *U.S. Patent No. 20100176066. Method of improving efficiency of UV photolysis of peracetic acid for disinfection and organic destruction.* Washington, DC: U.S. Patent and Trademark Office.

Buxton, G. V., Greenstock, C. L., Helman, W. P., & Ross, A. B. (1988). Critical review of rate constants for reactions of hydrated electrons, hydrogen atoms and hydroxyl radicals. *Journal of Physical and Chemical Reference Data,* 17, 513–886.

Cai, M., Sun, P., Zhang, L., & Huang, C. H. (2017). UV/peracetic acid for degradation of pharmaceuticals and reactive species evaluation. *Environmental Science and Technology, 51*(24), 14217–14224.

Caretti, C., & Lubello, C. (2003). Wastewater disinfection with PAA and UV combined treatment: a pilot plant study. *Water Research*, 37(10): 2365–2371. https://doi.org/10.1016/S0043-1354(03)00025-3

Chhetri, R. K., Bonnerup, A., & Andersen, H. R. (2016). Combined sewer overflow pretreatment with chemical coagulation and a particle settler for improved peracetic acid disinfection. *Journal of Industrial and Engineering Chemistry*, 37, 372–379.

Chhetri, R. K., Thornberg, D., Berner, J., Gramstad, R., Öjstedt, U., Sharma, A. K., & Andersen, H. R. (2014). Chemical disinfection of combined sewer overflow waters using performic acid or peracetic acids. *The Science of the Total Environment*, 490, 1065–1072.

Dell'Erba, A., Falsanisi, D., Liberti, L., Notarnicola, M., & Santoro, D. (2004). Disinfecting behaviour of peracetic acid for municipal wastewater reuse. *Desalination*, 168, 435–442.

Dietrich, J. P., Başağaoğlu, H., Loge, F. J., & Ginn, T. R. (2003). Preliminary assessment of transport processes influencing the penetration of chlorine into wastewater particles and the subsequent inactivation of particle-associated organisms. *Water Research, 37*(1), 139–149.

Dietrich, J. P., Loge, F. J., Ginn, T. R., & Başağaoğlu, H. (2007). Inactivation of particle-associated microorganisms in wastewater disinfection: Modeling of ozone and chlorine reactive diffusive transport in polydispersed suspensions. *Water Research, 41*(10), 2189–2201.

Domínguez Henao, L., Cascio, M., Turolla, A., & Antonelli, M. (2018). Effect of suspended solids on peracetic acid decay and bacterial inactivation kinetics: Experimental assessment and definition of predictive models. *The Science of the Total Environment*, 643, 936–945.

Domínguez Henao, L., Delli Compagni, R., Turolla, A., & Antonelli, M. (2018). Influence of inorganic and organic compounds on the decay of peracetic acid in wastewater disinfection. *Chemical Engineering Journal,* 337, 133–142. doi:10.1016/j.cej.2017.12.074.

Dunkin, N., Coulter, C., Weng, S., Jacangelo, J. G., & Schwab, K. J. (2018, November 15). Effects of pH variability on peracetic acid reduction of human norovirus GI, GII RNA, and infectivity plus RNA reduction of selected surrogates. *Food and Environmental Virology*.

Dunkin, N., Weng, S., Schwab, K. J., McQuarrie, J., Bell, K., & Jacangelo, J. G. (2017). Comparative inactivation of murine Norovirus and Ms2 bacteriophage by peracetic acid and monochloramine in municipal secondary wastewater effluent. *Environmental Science & Technology*, 51.5, 2972–2981.

Elouear, Z., Bouzid, J., Boujelben, N., Feki, M., Jamoussi, F., & Montiel, A. (2008). Heavy metal removal from aqueous solutions by activated phosphate rock. *Journal of Hazardous Materials*, *156*(1–3), 412–420.

European Centre for Ecotoxicology and Toxicology of Chemicals. (2001). Joint Assessment of Commodity Chemicals (JACC) programme. JACC No. 40—Peracetic Acid (CAS No. 79-21-0) and its Equilibrium Solutions. Brussels, Belgium.

Falsanisi, D., Gehr, R., Santoro, D., Dell, E. A., Notarnicola, M., & Liberti, L. (2006). Kinetics of PAA demand and its implications on disinfection of wastewaters. *Water Quality Research Journal of Canada*, *41*(4), 398–409.

Falsanisi, D., Gehr, R., Liberti, L., & Notarnicola, M. (2008). Effect of suspended particles on disinfection of a physicochemical municipal wastewater with peracetic acid. *Water Quality Research Journal of Canada*, *43*(1), 47–54.

Finnegan, M., Linley, E., Denyer, S. P., McDonnell, G., Simons, C., & Maillard, J. Y. (2010). Mode of action of hydrogen peroxide and other oxidizing agents: Differences between liquid and gas forms. *Journal of Antimicrobial Chemotherapy*, *65*(10), 2108–2115.

Fraser, J. A. L., Godfree, A. F., & Jones, F. (1984). Use of peracetic acid in operational sewage sludge disposal to pasture. *Water Science Technology*, 17, 451–466.

Gehr, R., & Cochrane, D. (2002). Peracetic acid (PAA) as a disinfectant for municipal wastewaters: Encouraging performance results from physicochemical as well as biological effluents. *Proceedings of the Water Environment Federation*, 1, 182–198.

Gonzalez, A., Gehr, R., Vaca, M., & Lopez, R. (2012). Disinfection of an advanced primary effluent with peracetic acid and ultraviolet combined treatment: A continuous-flow pilot plant study. *Water Environment Research*, *84*(3), 247–253.

Greenspan, F. P., & Gall, R. J. (1956, September 1). Epoxy fatty acid ester plasticizers. Preparation and properties. *Journal of the American Oil Chemists' Society*, *33*(9), 391–394.

Haas, C. N., & Finch, G. R. (2001). *Methodologies for the determination of disinfection effectiveness.* Denver, CO: American Water Works Association Research Foundation and American Water Works Association.

Howarth, J., & Harvey, M. (2010). *U.S. Patent No. US7651860B2. Method of analyzing low levels of peroxyacetic acid in water.* Washington, DC: U.S. Patent and Trademark Office.

Irani, R. R., & Morgenthaler, W. W. (1963). Iron sequestration by polyphosphates. *Journal of the American Oil Chemical Society, 40*(7), 283–285.

Jurkiewicz, C. H., Kunigk, L., & Silva, S. M. (2012, July 1). The influence of temperature and organic matter on the decomposition kinetics of peracetic acid in aqueous. *Latin American Applied Research, 42*(3), 291–297.

Kerkaert, B., Mestdagh, F., Cucu, T., Aedo, P. R., Ling, S. Y., & Meulenaer, B. (2011). Hypochlorous and peracetic acid induced oxidation of dairy proteins. *Journal of Agricultural Food Chemistry, 59*(3), 907–914.

Kitis, M. (2004). Disinfection of wastewater with peracetic acid: A review. *Environment International, 30*(1), 47–55.

Klueh, K. G., & Robinson, R. B. (1988). Sequestration of iron in groundwater by polyphosphates. *Journal of Environmental Engineering, 114*(5), 1192–1199.

Kobylinski, E. A., Hunter, G. L., & Shaw, A. R. (2006). On line control strategies for disinfection systems: Success and failure. *Proceedings of the 79th Annual Water Environment Federation Technical Exhibition and Conference* [CD-ROM]; Dallas, Texas, Oct 21–25; Alexandria, VA: Water Environment Federation.

Koivunen, J., & Heinonen-Tanski, H. (2005a). Peracetic acid (PAA) disinfection of primary, secondary and tertiary treated municipal wastewaters. *Water Research, 39*(18), 4445–4453.

Koivunen, J., & Heinonen–Tanski, H. (2005b). Inactivation of enteric microorganisms with chemical disinfectants, UV irradiation and combined chemical/UV treatments. *Water Research*, 39, 1519–1526.

Kunigk, L., & Almeida, M. C. B. (2001). Action of peracetic acid on *Escherichia coli* and Staphylococcus aureus in suspension or settled on stainless steel surfaces. *Brazilian J. Microbiol., 32*(1), 38–41.

Larson, T. E. (1957). Evaluation of use of polyphosphates in water industry. *Journal of the American Water Works Association, 49*(12).

Lazarova, V., Janex, M., Fiksdal, L., Oberg, C., Barcina, I., & Pommepuy, M. (1998). Advanced wastewater disinfection technologies: Short and long term efficiency. *Water Science and Technology, 38*(12), 109–117.

LeChevallier, M. W., Hassenauer, T. S., Camper, A. K., & McFeters, G. A. (1984). Disinfection of bacteria attached to granular activated carbon. *Applied and Environmental Microbiology*, *48*(5), 918–923.

Lefevre, F., Audic, J. M., & Ferrand, F. (1992). Peracetic acid disinfection of secondary effluents discharged off coastal seawater. *Water Science and Technology*, 25, 155–164.

Liberti, L., & Notarnicola, M. (1999). Advanced treatment and disinfection for municipal wastewater reuse in agriculture. *Water Science and Technology*, 40, 235–245.

Liu, D., Steinberg, C. E. W., Straus, D. L., Pedersen, L.-F., & Meinelt, T. (2014). Salinity, dissolved organic carbon and water hardness affect peracetic acid (PAA) degradation in aqueous solutions. *Aquacultural Engineering*, 60, 35–40.

Lubello, C., Caretti, C., & Gori, R. (2002). Comparison between PAA/UV and H2O2/UV disinfection for wastewater reuse. *Water Supply*, *2*(1), 205–212.

Luna-Pabello, V. M., Miranda Ríos, M., Jiménez, B., & Orta de Velasquez, M. T. (2009). Effectiveness of the use of Ag, Cu and PAA to disinfect municipal wastewater. *Environmental Technology*, *30*(2), 129–139.

Luukkonen, T., Heyninck, T., Rämö, J., & Lassi, U. (2015). Comparison of organic peracids in wastewater treatment: Disinfection, oxidation and corrosion. *Water Research*, 85, 275–285.

Luukkonen, T., & Pehkonen, S. O. (2017). Peracids in water treatment: A critical review. *Critical Reviews in Environmental Science and Technology*, *47*(1), 1–39.

Luukkonen, T., Teeriniemi, J., Prokkola, H., Rämö, J., & Lassi, U. (2014). Chemical aspects of peracetic acid based wastewater disinfection. *Water SA*, *40*(1), 73–80.

Mattle, M. J., Crouzy, B., Brennecke, M. R., Wigginton, K., Perona, P., & Kohn, T. (2011). Impact of virus aggregation on inactivation by peracetic acid and implications for other disinfectants. *Environmental Science & Technology*, *45*(18), 7710–7717.

McFadden, M., Loconsole, J., Schockling, A. J., Nerenberg, R., & Pavissich, J. P. (2017). Comparing peracetic acid and hypochlorite for disinfection of combined sewer overflows: Effects of suspended-solids and pH. *The Science of the Total Environment*, 599–600, 533–539.

Mezzanotte, V., Antonelli, M., Citterio, S., & Nurizzo, C. (2007). Wastewater disinfection alternatives: Chlorine, ozone, peracetic acid, and UV light. *Water Environment Research*, 79, 2373–2379.

Mir, J., Morató, J., & Ribas, F. (1997). Resistance to chlorine of freshwater bacterial strains. *Journal of Applied Microbiology*, *82*(1), 7–18.

Murray, A., Goldman, J., Sarathy, S. R., Hilts, B., Bell, K., Santoro, D., . . . Brower, M. (2016). Disinfection of a municipal wastewater secondary effluent with a combination of ultraviolet irradiation and peracetic acid. *Proceedings of the 89th Annual Water Environment Federation Technical Exhibition and Conference*; New Orleans, Louisiana, Sept 24–28; Alexandria, VA: Water Environment Federation.

Orta de Velásquez, M., Yáñez-Noguez, I., Jiménez-Cisneros, B., & Luna Pabello, V. M. (2017). Adding silver and copper to hydrogen peroxide and peracetic acid in the disinfection of an advanced primary treatment effluent. *Environmental Technology*, 29, 1209–1217.

Patnaik, P. (2008). *Handbook of environmental analysis: Chemical pollutants in air, water, soil, and solid wastes.* Boca Raton, FL: CRC Press, Taylor & Francis Group.

Pedersen, L. F., Meinelt, T., & Straus, D. L. (2013). Peracetic acid degradation in freshwater aquaculture systems and possible practical implications. *Aquacultural Engineering*, 53, 65–71.

PeroxyChem. (2014). Disinfection Digest No 2, June 2014.

PeroxyChem. (2015). Disinfection Digest No 10, October 2015.

Popov, E., Eloranta, J., Hietapelto, V., Vuorenpalo, V.-M., Aksela, R., & Jäkärä, J. (2005). Mechanism of decomposition of peracetic acid by manganese ions and diethylenetriaminepentaacetic acid (DTPA). *Holzforschung*, *59*(5), 507–513.

Rajala-Mustonen, R. L., Toivola, P. S., & Heinonen-Tanski, H. (1997). Effects of peracetic acid and UV irradiation on the inactivation of coliphages in wastewater. *Water Science and Technology*, 35, 237–241.

Richardson, S. D., Simmons, J. E., & Rice, G. (2002). Disinfection byproducts: The next generation. *Environmental Science & Technology*, 36, 198A–205A.

Rossi, S., Antonelli, M., Mezzanotte, V., & Nurizzo, C. (2007, January 1). Peracetic acid disinfection: A feasible alternative to wastewater chlorination. *Water Environment Research*, 41–50.

Rothbart, S., Ember, E. E., & van Eldik, R. (2012). Mechanistic studies on the oxidative degradation of orange II by peracetic acid catalyzed by simple manganese(ii) salts. Tuning the lifetime of the catalyst. *New Journal of Chemistry*, *36*(3), 732–748.

Sánchez-Ruiz, C., Martinez-Royano, S., & Tejero-Monzon, I. (1995). An evaluation of the effciency and impact of raw WW disinfection with PAA prior to ocean discharge. *Water Science and Technology*, *32*(7), 159–166.

Santoro, D., Gehr, R., Bartrand, T. A., Liberti, L., Notarnicola, M., Dell'Erba, A., Falsanisi, D., . . . Haas, C. N. (2007, July 1). Wastewater disinfection by peracetic acid: Assessment of models for tracking residual measurements and inactivation. *Water Environment Research, 79*(7), 775–787.

Santoro, D., Crapulli, F., Raisee, M., Raspa, G., & Haas, C. N. (2015, January 1). Nondeterministic computational fluid dynamics modeling of Escherichia coli inactivation by peracetic acid in municipal wastewater contact tanks. *Environmental Science & Technology, 49*(12), 7265–75.

Sarathy, S., Santoro, D., Gaida, D., Canali, A., Turolla, A., Antonelli, M., . . . Neofotistos, P. (2017). This is how we do it—Dose based control for wastewater disinfection with peracetic acid. *Proceedings of the 90th Annual Water Environment Federation Technical Exhibition and Conference*; Chicago, Illinois, Sept 30–Oct 4; Alexandria, VA: Water Environment Federation.

Soares Cavallini, G., De Campos, S. X., Beber De Souza, J., & De Sousa Vidal, C. M. (2013). Evaluation of the physical-chemical characteristics of wastewater after disinfection with peracetic acid. *Water, Air & Soil Pollution, 224*(10).

Stampi, S., De Luca, G., & Zanetti, F. (2001). Evaluation of the efficiency of peracetic acid in the disinfection of sewage effluents.*Journal of Applied Microbiology*, 91, 833–838.

Tchobanoglous, G., Stensel, H. D., Tsuchihashi, R., Franklin, B., Abu-Orf, M., Bowden, G., & Pfrang, W. (2014). *Wastewater engineering: Treatment and resource recovery* (5th ed.). New York, NY: Metcalf and Eddy, Inc., McGraw-Hill Book Company.

Turolla, A., Sabatino, R., Fontaneto, D., Eckert E. M., Colinas, N., Corno, G., Citterio, B., Biavasco, F., Antonelli, M., Mauro, A., Mangiaterra, G., Di Cesare, A. (2017). Defence strategies and antibiotic resistance gene abundance in enterococci under stress by exposure to low doses of peracetic acid. *Chemosphere*, 185, 480–488.

Varner, D. L., Skipton, S., Hay, D., & Jasa, P. J. (1996). G96-1280 drinking water: Iron and manganese. *Historical Materials from University of Nebraska-Lincoln Extension.*

Vianello, C., Salzano, E., & Maschio, G. (2018, January 1). Thermal behaviour of peracetic acid for the epoxydation of vegetable oils in the presence of catalyst. *Process Safety and Environmental Protection.*

Wagner, M., Brumelis, D., & Gehr, R. (2002). Disinfection of wastewater by hydrogen peroxide or peracetic acid: Development of procedures for measurement of residual disinfectant and application to a physicochemically treated municipal effluent. *Water Environment Research*, 74, 33–50.

Winward, G. P., Avery, L. M., Stephenson, T., & Jefferson, B. (2008). Chlorine disinfection of grey water for reuse: Effect of organics and particles. *Water Research*, *42*(1–2), 483–491.

Yuan, Z., Ni, Y., & Van Heinegen, A. R. P. (1997a). Kinetics of peracetic acid decomposition: Part I: Spontaneous decomposition at typical pulp bleaching conditions. *Canadian Journal of Chemical Engineering*, *75*(1), 37–41.

Yuan, Z., Ni, Y., & Van Heiningen, A. R. P. (1997b). Kinetics of the peracetic acid decomposition: Part II: pH effect and alkaline hydrolysis. *Canadian Journal of Chemical Engineering*, 75, 42–47.

Zhang, C.; Brown, P. J. B., & Hu, Z. (2018). Thermodynamic properties of an emerging chemical disinfectant, peracetic acid. *The Science of the Total Environment,* 621, 948–959.

Zhao, X., Cheng, K., Hao, J., & Liu, D. (2008, April 1). Preparation of peracetic acid from hydrogen peroxide, part II: Kinetics for spontaneous decomposition of peracetic acid in the liquid phase. *Journal of Molecular Catalysis A: Chemical*, 284, 58–68.

5.0 SUGGESTED READING

Pedersen, L. F., Pedersen, P. B., Nielsen, J. L., & Nielsen, P. H. (2009). Peracetic acid degradation and effects on nitrification in recirculating aquaculture systems. *Aquaculture*, *296*(3–4), 246–254.

4

Peracetic Acid Analysis and Process Monitoring Tools

Allegra da Silva, Joanne Carpenter, Nicole McLellan, Nicole Stephens, Elisabeth Werth, & Ian Myers

1.0 INTRODUCTION TO PERACETIC ACID ANALYSIS

Several methods of analysis for the measurement of peracetic acid (PAA) have been reported in the literature dating back to the 1910s (Luukkonen & Pehkonen, 2017). These represent various modes of analytical techniques: titrimetric (redox), colorimetric, enzymatic, amperometric, and chromatography. Some methods do not lend themselves well for use in the field. Because this publication is a practical guide to implementing PAA disinfection at water resource recovery facilities (WRRFs), emphasis has been given to methodologies with operator and technician personnel in mind. Table 4.1 provides a summary of PAA measurement techniques and currently available instruments.

2.0 DEVELOPING OFFICIALLY APPROVED METHODS

2.1 U.S. Environmental Protection Agency 40 C.F.R. 136

At the time of this writing (2020), there are no approved methods for determining PAA residual in wastewater published in the *Federal Register* [40 C.F.R. Part 136 Guidelines Establishing Test Procedures for the Analysis of Pollutants]. This hinders PAA disinfection implementation efforts and results in a lack of uniformity in the way that regional and state environmental agencies (responsible for issuing permits to publicly owned treatment works) address PAA monitoring requirements. For a new method to be approved and published in the *Federal Register*, it must be subjected to rigorous validation protocols that evaluate method sensitivity, matrix interference effects, and precision and accuracy. Consensus-standard organizations such as ASTM International or *Standard Methods for the Examination of Water and Wastewater* (American Public Health Association, 2012) can lead these efforts. However, in order for the method to be published in one of the tables of approved test procedures in 40 C.F.R. Part 136, it must undergo legislative rulemaking. The U.S. Environmental Protection Agency (U.S. EPA) issues updates to the Clean Water Act on an irregular basis; therefore, it may take 3 to 5 years before the method is published to work through the various elements of the legislative process (including U.S. EPA review, a public comment period, and U.S. EPA's rebuttal to those comments).

2.2 Standard Methods for the Examination of Water and Wastewater

In 2018, an effort was underway under the auspices of *Standard Methods* to establish a new method for the analysis of PAA based on the DPD method

TABLE 4.1 Summary of PAA measurement techniques (compiled from vendor information as of February 2020).

Method	Description	Application	Instruments	Detection range	Accuracy
Colorimetric	Concentration determined by DPD[e] reaction with sample and potassium iodide (KI) in excess. Intensity of resulting pink color (as measured by color chart or calibrated photometer) is directly proportional to PAA concentration.	Grab sampling in field for process control or compliance measurements[b]. Bench/pilot-scale in-stream concentration measurement. Verification of online meters. Hach CL17 applicable for continuous, in situ residual measurement for monitoring or process control.	CHEMetrics I2020	0–1 ppm (visual test kit)	+/− 20% at low end of detection limit
			Single Analyte Meter (SAM), V-2000 or V-3000 multiparameter photometer	0–5 ppm	=/− 30% at detection limit
			Hach Company CL17	0–5 ppm	+/− 5%[c]
			Hach Company MR and HR Chlorine Pocket Colorimeter™ (PCII), DR 900 multiparameter photometer	0.1–10 ppm	+/− 4%
			Masters Company, Inc. MP-9700 PAA Test Kit, Dual Range	0.1–10 ppm	
Amperometric	Voltage applied across electrodes (cathode and anode). Concentration is determined by change in current that occurs when PAA is reduced at the cathode.	Continuous, in situ residual measurement for monitoring or process control.	Prominent DULCOTEST® Sensors	0–200 ppm 0–2000 ppm	
			ATI Q46/85 Peracetic Acid Monitor	0–2 ppm 0–200 ppm	+/− 2%[c]
			Endress + Hauser CCS120	0.1–10 ppm	
Titrimetric	Concentration determined by titration endpoint as indicated by change in color (when KI added in excess) or by electric current measured amperometrically.	Validation of bulk chemical strength. Bench/pilot-scale in-stream concentration measurement[a]. Compliance sampling[a, b].	Masters Company Inc. #D-10015-02	5–150 ppm	
Enzymatic	Catalase is used to selectively eliminate hydrogen peroxide from the PAA sample before either titrimetric or colorimetric measurement.	Grab sampling in field for process control or compliance measurements. Bench/pilot-scale in-stream concentration measurement.	N/A	0–10 ppm	+/− 10 − 30%[d]

[a] In laboratory only. Titration kits are not appropriate for low, in-stream concentration measurement.

[b] Accuracy decreases as time between sample extraction and test increases. Titrations are not feasible for compliance sampling if testing station is not within reasonable range of sampling location.

[c] Accuracy dependent on quality of sample delivered to instrument. Assumes zero demand in sample system upstream of instrument.

[d] Accuracy dependent on specific determinative method and PAA concentration level.

[e] DPD refers to *N-N*-diethyl-*p*-phenylendiamine.

for publication in Section 4000. (DPD stands for the chemical compound *N*-*N*-diethyl-*p*-phenylendiamine.) The *Standard Methods* process for validating new methods relies on the establishment of a joint task group (JTG). The JTG is charged with writing the method and developing validation protocols, which include defining the method scope, purpose, applicability, and performance before commencing development or validation work. The method validation process demonstrates that the analytical method is suitable for its intended use.

Validation involves conducting a variety of studies to evaluate method performance under defined conditions. *Standard Methods* classifies methods in two categories: proposed and standard. Proposed means the method has undergone development and validation that meets the requirements set forth in Section 1040 of *Standard Methods*. To meet the standard classification, a method must also be widely used.

The goal of the validation procedure is to ensure that the analytical results of the method are scientifically sound and legally defensible. Method development is a four-step process that involves the following:

- Determination of single operator precision and bias,
- Analysis of various applicable matrices,
- Determination of method ruggedness, and
- Collaborative study.

3.0 ANALYSIS TECHNIQUES

3.1 Colorimetric (DPD Method)

3.1.1 Method History

The DPD method of analysis has been the most widely used procedure for measuring chlorine and other oxidizing disinfectants at drinking water treatment plants and WRRFs for decades (Palin, 1986). It has withstood the test of time because it is an inexpensive, sensitive, rugged, accurate, fast, and straightforward method that can be successfully used by operator- or technician-level employees. For these reasons, the DPD method (in particular, the colorimetric version) has found use as a routine method of measurement for PAA in wastewater final effluent (Graham, Lomax, & Glabra, 2012; Luukkonen et al., 2014; PeroxyChem, 2014; Wilson, 2014).

Dr. Arthur T. Palin was the first to publish a paper about the DPD method for the measurement of free chlorine and chlorine combined with ammonia (chloramines) in water (Palin, 1957). Palin recognized there was a need for a simple and reliable method for the routine measurement of free and combined residual chlorine that could be carried out by water treatment operators. He partnered with a company called Wilkinson & Simpson Ltd., now known as Palintest Ltd., to develop a formulation for a DPD reagent in tablet form. The test kit was marketed for disinfection process monitoring in municipal water treatment plants, swimming pools, wastewater effluents, and industrial water treatment.

In 1960, Wilkinson & Simpson Ltd. was granted an exclusive license to manufacture and market the DPD tablet. Eventually, as the number of DPD manufacturers increased, the form of the DPD reagent expanded to include powders packaged in foil packets, liquids, or powders packaged in self-filling vacuum-sealed glass ampoules and test strips. The method has been approved by U.S. EPA as Method 330.5 and is published in *Standard Methods,* Part 4500–Cl.F and G.

The DPD method is also commonly used as a means to measure other oxidizing disinfection agents including bromine, chlorine dioxide, ozone, and peracetic acid (PAA). Historically, these oxidizing disinfectants have not been commercially available as stable standards; thus, analysts have been unable to generate analyte-specific calibrations and, therefore, commonly make the appropriate molecular weight adjustment to chlorine results to obtain test results that are specific to the disinfectant being evaluated.

3.1.2 Generalized Introduction to Technique

The DPD method is based on the reaction of low concentration levels (typically ≤5 ppm) of disinfecting agents with DPD at a reaction pH of 6.2 to 6.5. The DPD reagents are formulated with a phosphate buffer that will bring the reaction pH to this region, as well as a sequestering agent, typically ethylene diamine tetra acetic acid (EDTA), to tie up reactive metals such as copper. Palin also suggested that EDTA promoted DPD stability by suppressing the oxidative effect of dissolved oxygen.

When the DPD amine is oxidized, two oxidation products are possible (Harp, 1995). The primary oxidation product is a semiquinone known as a *Würster dye.* This comparatively stable free radical species produces the pink color in the DPD test. The secondary DPD oxidation product is an unstable, colorless imine compound. This form is favored under higher oxidant levels. Therefore, chlorine itself and other oxidants at concentrations significantly above the test range may prevent proper color development, causing a false low result.

Some disinfection agents directly oxidize DPD to a pink color (e.g., chlorine and chlorine dioxide), whereas others (e.g., chloramines, hydrogen peroxide, ozone, and PAA) develop little or no color directly with DPD.

These disinfection agents require the addition of iodide to the sample. Tablets and foil-pack DPD reagents intended for use for the measurement of total chlorine (free chlorine plus chloramines) are formulated with potassium iodide, whereas liquid DPD reagents supply the iodide as a separate accessory solution, which is added to the sample before the DPD. With the exception of hydrogen peroxide, these disinfection agents quickly oxidize iodide to form iodine as the triiodide ion (I_3^-). The reaction of iodide with PAA is given as follows (Equation 4.1):

$$CH_3COOOH + 3I^- + 2H^+ \rightarrow CH_3COOH + I_3^- + H_2O \qquad (4.1)$$

Triiodide, in turn, reacts with DPD, forming the Würster dye oxidation product. The DPD methods for hydrogen peroxide are modified further by the addition of an acidic molybdate solution or reagent to catalyze the hydrogen peroxide–iodide reaction producing triiodide (Parker, 1978).

Two DPD methods of analysis are available: titrimetric and colorimetric. In the titrimetric method, DPD functions as an indicator. If the disinfectant agent is present, the sample will turn pink after the addition of the DPD reagent. The sample is then titrated with standardized ferrous ammonium sulfate to a colorless endpoint reaction. If total chlorine, ozone, or PAA are being measured, then the sample is also treated with iodide.

3.1.3 Colorimetric Detection Options

In the colorimetric DPD version, the intensity of developed color is proportional to disinfectant concentration. This permits either visual or instrumental detection. The DPD colorimetric test kits are widely available. Visual test kits supply color standards with intensities and hue that correspond to a specific concentration of disinfectant. These "color comparators" take many forms (e.g., color wheels, charts, glass chips, and liquid dyes). They all rely on the analyst's visual comparison of the developed color in their sample with the color standards. Instrumental test kits are based on the absorption spectrum of the Würster dye, which is characterized by a double peak that can be measured photometrically at wavelengths ranging from 490 to 555 nm (Figure 4.1). Manufacturers develop calibrations for spectrophotometers and photometers that are programmed to display the test result. Many organizations require instrumental detection to eliminate concerns about subjectivity of visual color comparison.

3.1.4 DPD Interferences and Reaction Time

Any oxidant that is capable of oxidizing DPD will interfere; this could include free and combined residual chlorine. Furthermore, any oxidant that

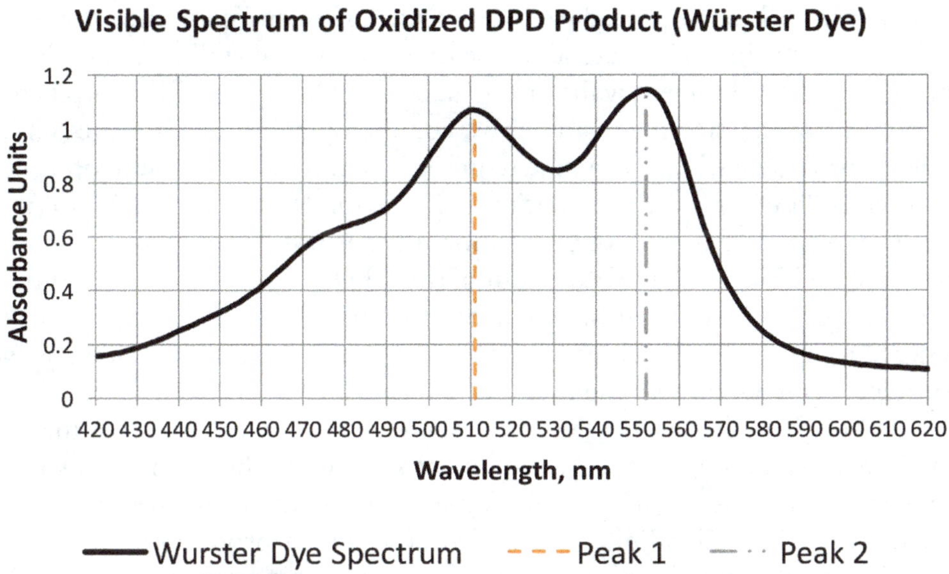

FIGURE 4.1 Visible spectrum of oxidized DPD product (Würster dye).

is capable of converting iodide to iodine will interfere with the test, including chloramines. However, because PAA wastewater disinfection is intended to replace chlorination, this should mean concerns about false positive PAA readings caused by chlorine are unwarranted. Knowledge about the treatment process of the system is needed to evaluate test results.

As with any colorimetric test, optical interferences stemming from sample background color and turbidity will be problematic. Steps to compensate for their effects can be used. For example, in visual tests, a color wheel comparator is designed to compare two tubes containing the water sample. One tube has not been treated with the colorimetric reagent(s) and serves as the "blank", while the water in the other tube is subjected to the color forming reagent. Both tubes are inserted in the color wheel compartment. As the color wheel is rotated, the reference color printed on the wheel is aligned with the blank tube. The analyst compares the developed color of the sample treated with the reagents to the blank tube and stops turning the wheel when a color intensity match has been made. In instrumental testing, the sample itself may be used to "zero" the instrument.

Alternatively, if the sample is suspected of containing substances that oxidize DPD or iodide before the addition of the disinfectant, a sample collected upstream from the disinfection contact chamber can be analyzed with DPD as a "reagent blank". The test result for the reagent blank can then be subtracted from the result obtained from the sample containing disinfectant.

The DPD test procedure for measuring PAA residual (i.e., the PAA remaining after oxidant demand has been met) in wastewater effluent is nearly identical to the procedure for measuring free and combined residual chlorine. The principal difference is the time required for full color development. Color development wait time must be fixed before color development resulting from interfering compounds, commonly called "break-through". The measurement of dichloramine requires 2 minutes for color development (American Public Health Association, 2012; Palin, 1957). In the DPD PAA test procedure, the pink developed color is either visually or instrumentally measured within 1 minute after adding the DPD reagent and iodide to the sample. For most applications, hydrogen peroxide break-through is considered to be negligible within 1 minute because hydrogen peroxide reacts extremely slowly with iodide (Sully & Williams, 1962).

Although hydrogen peroxide break-through is not a problem in the DPD method as long as the measurement is made within 1 minute, some researches have used catalase as a sample pretreatment before measurement with DPD. Cavallini et al. (2013) described a procedure where the sample was dosed with catalase and the various constituents that comprise DPD reagent: EDTA, phosphate buffer, DPD, and potassium iodide (KI). No details were provided about the catalase incubation time. Developed color was measured at 530 nm. This method was recommended for PAA measurement in the range 0.1 to 10 mg/L.

Binder and Menger (2000) used both forms of enzymes. The color-forming reagent used in their combinatorial in situ peracid generation research was Azino-bis(3-ethylbenzothiazoline-6-sulfonate) diammonium salt (ABTS). The samples were incubated 10 minutes with catalase first to decompose hydrogen peroxide (Equation 4.2). The peracid was then treated with ABTS in the presence of peroxidase. Peroxidase promotes the stoichometric oxidation of ABTS. In its oxidized form, ABTS produces a green color that is measured photometrically at 405 nm (Equation 4.3).

$$2H_2O_2 + RC(O)OOH \xrightarrow[\text{10 minutes}]{\text{catalase, pH} = 7.0} 2H_2O + O_2 + RC(O)OOH \qquad (4.2)$$

$$ABTS_{(red)} + RC(O)OOH \xrightarrow{\text{peroxidase, pH} = 5.0} ABTS_{(ox)} + RC(O)OH \qquad (4.3)$$

Lighting conditions should be controlled to minimize direct light exposure, which is suspected of causing ABTS photooxidation and/or light-mediated enzyme deactivation. Lastly, the authors reported that inconsistent results were obtained when hydrogen peroxide levels exceeded a 200-fold molar excess over the peracid.

3.1.5 Hydrogen Peroxide Analysis in the Presence of Peracetic Acid

Some environmental regulators may want to learn more about site-specific hydrogen peroxide decomposition rates (PeroxyChem, 2015). For this reason, they may require hydrogen peroxide monitoring during PAA disinfection validation trials and/or on an ongoing basis. Peracetic acid causes a false positive interference with colorimetric hydrogen peroxide test methods.

One PAA DPD test kit vendor (CHEMetrics, Inc., 2020) offers a technical bulletin that describes a procedure in which the sample is analyzed with their hydrogen peroxide test kit in the presence of PAA. The kit uses a ferric thiocyanate colorimetric method, in which the hydrogen peroxide oxidizes ferrous iron to ferric iron, forming an orange-colored complex with the thiocyanate ion. Peracetic acid will interfere with the test unless the sample is pre-treated with potassium iodide, converting the PAA to acetic acid and producing iodine. Neither acetic acid nor iodine interferes with this method. Hence, it is possible to obtain hydrogen peroxide and PAA test results quickly and easily in the field.

3.1.6 Example of Commercially Available Online Colorimetric Probe

Hach CL17 is applicable for continuous, in situ residual measurement for monitoring or process control.

3.2 Amperometric Methods

3.2.1 Generalized Introduction to Technique

Commercially available online amperometric sensors are commonly used for continuous measurement of PAA in wastewater. The typical amperometric sensor design applies DC voltage to two electrodes: a reference/counter silver/silver chloride and a gold cathode also known as a *collector* or *working electrode*. When PAA is reduced at the cathode, a change in current is detected. The current is converted to a temperature-compensated output signal that is proportional to the PAA concentration. Flowrate and pressure must be carefully controlled for accurate measurement. The sensing electrodes are isolated from the sample by a permeable diffusion membrane or diaphragm. A cap filled with electrolyte encloses the diaphragm. Care must be taken to exclude air bubbles when filling the cap with electrolyte. Bubbles cause erroneous readings. Power to the sensors is supplied externally from a measuring and control system.

3.2.2 Examples of Commercially Available Online Probes

ProMinent® and Analytical Technology, Inc., supply PAA sensors, controllers, and metering pumps (Figure 4.2). The system must be allowed to

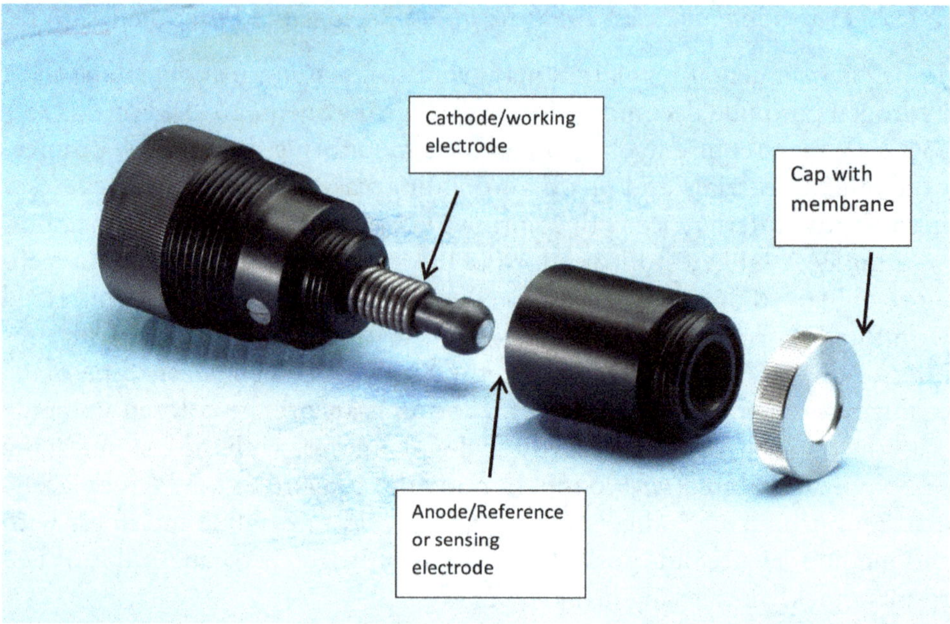

FIGURE 4.2 Close-up of Analytical Technology, Inc., PAA Model Q45/85 Amperometric Sensor (reprinted with permission from Analytical Technology, Inc.).

reach steady state during initial commissioning or while recommissioning or changing the diaphragm. This can take 1 to 2 hours. The sensor calibration should also be validated under these situations as well. Calibration involves measuring the PAA using a reference method. Typically, commercially available DPD test kits are used for this purpose (refer to Section 3.1). Alternatively, a dual titration can be performed (refer to Section 3.3). Periodic maintenance is required to keep the system free from deposits and air bubbles. The diaphragm cap must be periodically wiped with a damp cloth. Discharge (8 to 30 gal/hr) from the sensor is sent to an open drain.

3.3 Titrimetric Methods

3.3.1 Generalized Introduction to Technique

Redox titrations are commonly used to standardize percent (%) levels of compounds that can either be oxidized or reduced. Redox reactions are comprised of two parts that always occur together: a reduced half and an oxidized half. The reduced half gains electrons and the oxidation number decreases, whereas the oxidized half loses electrons and the oxidation number increases. The analyte being measured is titrated against a titrant that has been standardized. Either visual indicators, such as ferroin, or a potentiometer, such as a platinum electrode, are used to determine the endpoint of the titration. Indicators change colors over a certain potential range. Once the concentration of analyte is established via titration, the concentration values for working standards (serial dilutions of the stock solution) are calculated from this value taking the dilution factor into account.

3.3.2 Biphasic Ceric Sulfate/Thiosulfate Titration

Because PAA is in equilibrium with hydrogen peroxide, the biphasic titration procedure is used to determine the concentration of both constituents. The PAA sample is acidified first with a sulfuric acid solution. Hydrogen peroxide is titrated first against standardized ceric sulfate (PAA does not titrate with ceric sulfate); it is oxidized to hydrogen and oxygen (refer to Equation 4.4). Ferroin is used as the indicator because its transition potential corresponds closely to the equivalence point potential of the hydrogen peroxide with ceric sulfate. It changes from orange to blue. The sample is then treated with potassium iodide. Peracetic acid oxidizes the iodide to iodine. The iodine is then titrated against standardized sodium thiosulfate using starch as the indicator (refer to Equations 4.5 and 4.6), as follows:

$$2Ce^{4+} + H_2O_2 \rightarrow 2Ce^{3+} + 2H^+ + O_2 \tag{4.4}$$

$$CH_3COOOH + 2I^- + 2H^+ \rightarrow CH_3COOH + I_2 + H_2O \tag{4.5}$$

$$I_2 + 2S_2O_3^{2-} \rightarrow 2I^- + S_4O_6^{2-} \tag{4.6}$$

One concern about the biphasic titration reported in the literature is that the acidic conditions of the titration will promote the hydrolysis of PAA, causing the concentration to change while performing the titration. To minimize this effect, the titration is carried out at a reduced temperature, typically by chilling the sulfuric acid solution below 10 °C (50 °F).

Cavallini et al. (2013) evaluated an iodometric titration of PAA, which is essentially the same method\reaction described in Equation 4.3, except the ceric sulfate titration with hydrogen peroxide is eliminated. A sample is pretreated with a potassium phosphate buffer and catalase and then homogenized for 60 seconds. The sample is then dosed with a mixed solution of sulfuric acid and ammonium molybdate. (Note: Molybdate catalyzes the reaction of hydrogen peroxide with potassium iodide. Because catalase eliminates the hydrogen peroxide from the sample, molybdate should be unnecessary. The authors do not explain why they included it in the assay.)

After potassium iodide is added to the sample it is titrated against sodium thiosulfate using starch as the indicator. Results are calculated using Equation 4.7, as follows:

$$\text{PAA mg/L} = (V_T \times 38\,000 \times 0.01)/V_S \qquad (4.7)$$

Where

V_T = the volume in mL of 0.01N sodium thiosulfate titrated and
V_S = the volume in mL of sample titrated.

According to Cavallini et al. (2013), this method produced low-biased PAA test results in a comparative study involving PAA standards prepared in deionized water. This was attributed to catalase degradation of PAA. Despite this shortcoming, the author concluded this method was suitable for the measurement of PAA in the range of 1 to 5 mg/L.

3.3.3 Timed Thiosulfate Repetitive Titration with Extrapolation

In this iodometric titration procedure, only sodium thiosulfate titrant is used to determine both PAA and hydrogen peroxide. Because the titration is carried out under milder acidic conditions than the biphasic titration (acetic acid rather than sulfuric acid), the risk of PAA hydrolysis is minimized. Once again, the PAA sample is pretreated with a solution of potassium iodide using starch as the indicator. Peracetic acid oxidizes iodide to iodine, which is titrated with sodium thiosulfate. Although hydrogen peroxide will also oxidize iodide to iodine, the reaction occurs at a much slower rate. This method incorporates a correction for the slow reaction between hydrogen peroxide and iodide. A molybdate catalyst is subsequently added to the sample, which catalyzes the reaction of hydrogen peroxide with iodide, and the titration with sodium thiosulfate continues.

The procedure involves timing the titration and documenting the volume of titrant at three points during the titration. To begin, chilled 0.1N acetic acid solution is added to an accurately weighed sample. The indicator is added to the sample. A timer is started as soon as 10 mL of a 15% w/v potassium iodide solution is added to the sample. The sample is titrated with 0.1N sodium thiosulfate titrant until the blue color is discharged. As soon as the blue color returns, the sodium thiosulfate is added dropwise until 2 minutes have elapsed from the addition of the iodide.

As soon as the blue color returns the second time, the burette volume is recorded (v_1) and the time is recorded (t_1). The dropwise titration resumes for another 3 or 4 minutes when a second pair of titrant volume (v_2) and corresponding time (t_2) readings are recorded when the blue color appears a third time. Then, three drops of a 5% ammonium molybdate solution are added to the sample. The titration continues until the colorless endpoint holds for at least 1 minute. The volume of titrant is recorded (v_3).

A corrected titer for PAA at time 0 (v_0) is calculated from the data as follows:

$$v_0 = v_1 - \frac{t_1(v_2 - v_1)}{(t_2 - t_1)} \tag{4.8}$$

To calculate PAA concentration (%):

$$\%\ \text{PAA} = \frac{v_0 \times \text{sodium thiosulfate normality} \times 38.03}{10\ \text{g}} \tag{4.9}$$

where

g = mass of sample.

The difference between this and the total titer (v_3) corresponds to the hydrogen peroxide concentration.

$$\%\ \text{H}_2\text{O}_2 = \frac{(v_3 - v_0) \times \text{sodium thiosulfate normality} \times 17.01}{10\ \text{g}} \tag{4.10}$$

where

g = mass of sample.

3.3.4 Acid Base Titration

The concentration of PAA and acetic acid (HAc) can be determined simultaneously using an acid base titration that takes advantage of the different pK_a values of the two acids. The titration lends itself to automated analysis using an auto titrator. Peracetic acid is a weaker acid than acetic acid, so a potentiometric titration can separate the two equivalence points. This is demonstrated when a plot of the titration curve is overlaid with the derivative curve, ΔpH/ΔmL. The pK_as for both acids can be read off the graph as the pH value that corresponds to the minimum ΔpH/ΔmL. Sode (2014) demonstrated that his data agreed very well with the published pK_a values of 4.75 for acetic acid and 8.2 for peracetic acid.

Sode (2014) provided further validation for this method of analysis by considering the effects of sample dilution, varying concentrations of hydrogen peroxide and HAc on PAA and HAc % recovery values, and evaluating the effect of elapsed time after sample dilution on test results. Sode concluded that the method is robust, and he promotes it as an alternative to the dual cerimetric/iodometric titration because it does not need to be conducted under reduced temperature conditions and does not require the addition of sulfuric acid. Of course, this method does not address the measurement of hydrogen peroxide.

3.4 Enzymatic Methods

Enzymes can be used to eliminate hydrogen peroxide before PAA measurement. Two forms of enzymes are available: catalase and peroxidase. Both are heme enzymes that catalyze reactions involving hydrogen peroxide. Catalase acts directly on hydrogen peroxide and promotes its decomposition to water and oxygen (Equation 4.11), as follows:

$$2H_2O_2 \xrightarrow{\text{catalase}} 2H_2O + O_2 \tag{4.11}$$

Peroxidase reacts by mechanisms similar to catalase; however, peroxidase facilitates hydrogen peroxide oxidation reaction on various substrates (Equation 4.12), as follows:

$$H_2O_2 + H_2R \xrightarrow{\text{peroxidase}} 2H_2O + R \tag{4.12}$$

The enzymes are commercially available from chemical supply companies. Lyophilized catalase and peroxidase are purified from bovine liver and the horseradish plant, respectively. The PAA literature cites methods that use both forms of enzymes, which find use in titrimetric and colorimetric methods.

4.0 FIELD EXPERIENCE WITH PERACETIC ACID MEASUREMENT

4.1 Peracetic Acid Residual Grab Sampling

4.1.1 Choosing a DPD Platform

Accurate measurement of PAA residual is essential to successful implementation of a PAA disinfection system. There are several established analytical techniques for determining PAA that were discussed in detail earlier in this chapter. Although all of the techniques can accurately measure PAA, the following questions should be considered to help determine which analytical technique is most appropriate for each application:

- What will the data be used for?
- Who will be performing the analysis and where?

- How frequently will sampling be required and how time consuming will analysis be?
- What level of staff training is needed for accurate results?
- Do the properties of the analyte introduce any special circumstances for the analysis?

At a WRRF, grab measurements will be used to verify online instrumentation readings, ensure proper dosing or quenching, and monitor effluent residual levels for permit compliance. These tasks require frequent sampling and analysis by a variety of staff, at all times of the day, and in every type of weather. Peracetic acid decomposes readily in wastewater effluent, necessitating an analysis with little setup and a quick reaction time.

When applied to monitoring treated wastewater effluent, these considerations recommend an analysis that is easy to learn and quick to perform, but also produces accurate results outside of an ideal laboratory environment. The DPD method addresses each of the requirements and, as an additional benefit, many operators are already familiar with the DPD method because it is an approved method to measure chlorine residual.

Several vendors offer DPD test kits and handheld photometers. The DPD reagents are available in pre-dosed glass ampoules, pre-measured powder packets, or tablets to be added to sample cells or they can be made up in bulk in a laboratory and measured out for each analysis. Some photometers have a pre-programmed calibration for PAA, whereas others are calibrated for chlorine and require a conversion. For photometers that display results in ppm Cl_2, the results must be converted to ppm PAA by multiplying by 1.07, which is the molecular weight ratio of PAA:Cl_2.

The choice between different test kit brands involves more than personal preference. Consideration should also be given to the materials and supplies needed for each analysis, glass disposal, equipment cleaning, photometer maintenance, and cost. Each photometer should also be evaluated for accuracy, precision, ease of use, and ruggedness.

Two test kits with their photometers made by different suppliers are compared in Figure 4.3. A standard method detection limit (MDL) study was run on each photometer in the laboratory using deionized water spiked with a known concentration of PAA. Because these photometers will be most often used in the field for analyzing PAA in treated wastewater, the MDL determination should be repeated under those conditions to better represent the realistic detection level. Unlike many analytes, PAA decomposes too quickly in treated wastewater to perform the field MDL study in that matrix. Instead, the same MDL study was run, using spiked deionized (DI) water, in the field to establish that the method has comparable performance

MDL Test Kit 1 - PAA Photometer			
Replicate	TRUE Value	LAB Result	FIELD Result
1		0.38	0.37
2		0.34	0.33
3		0.39	0.43
4	0.40	0.36	0.41
5		0.39	0.35
6		0.39	0.34
7		0.32	0.45
n		7	7
Mean		0.37	0.38
Standard Deviation		0.03	0.05
T-value (98% CL, 6 DF)		3.143	3.143
MDL		**0.09**	**0.15**

Test Kit 1 - PAA Photometer			
Replicate	TRUE Value	Result	% Recovery
1		2.54	101.6
2	2.50	2.45	98.0
3		2.47	98.8
4		2.47	98.8
Mean		2.48	99.3
Acceptable Range		2.00 - 3.00	80 - 120
Method Acceptance		**PASS**	

MDL Test Kit 2 - Cl_2 Photometer*1.07			
Replicate	TRUE Value	LAB Result	FIELD Result
1		0.49	0.51
2		0.51	0.52
3		0.54	0.49
4	0.40	0.47	0.50
5		0.52	0.51
6		0.55	0.50
7		0.49	0.47
n		7	7
Mean		0.51	0.50
Standard Deviation		0.03	0.02
T-value (98% CL, 6 DF)		3.143	3.143
MDL		**0.08**	**0.05**

Test Kit 2 - Cl_2 Photometer*1.07			
Replicate	TRUE Value	Result	% Recovery
1		1.27	127.3
2	1.00	1.25	125.2
3		1.26	126.3
4		1.26	126.3
Mean		1.26	126.3
Acceptable Range		0.80 - 1.20	80 - 120
Method Acceptance		**FAIL**	

FIGURE 4.3 Example of verification data for two types of test kits used in selecting preferred platform.

in the environment where it will be used most often. Both test kits performed similarly in each location.

Another easy step to evaluate the capability of each photometer is to look at the percent recovery. Deionized water was spiked with a known concentration of PAA in the middle of each test kits' stated measurement range. Looking at the percent recoveries from each photometer shows a high result bias in Test Kit 2. Because the results and percent recoveries fell outside of the acceptable range, the photometer failed its initial demonstration of capability. Test Kit 1 was chosen for all process control and compliance sampling for this application.

4.1.2 Accurate Sampling

After validating the performance of test kits, the next concern to address is sampling because an accurate result cannot be obtained from a poor sample.

Many of the properties that make PAA an attractive disinfectant, such as its quick decomposition in wastewater effluent, also make it difficult to measure and limit sample storage time.

Grab samples should be collected in a container with a tightly fitting lid. The container should be filled and capped, avoiding sample agitation. The sample location should be chosen from a well-mixed area, avoiding stagnant corners or eddies pools. The location should be well-marked to ensure samples are consistent from operator to operator. The analysis should begin immediately upon collection, and the number of samples to be analyzed should be kept to an absolute minimum. Figure 4.4 emphasizes this point by summarizing a study on the rate of two different initial concentrations of PAA and their decay in wastewater effluent compared to deionized water. A vertical line marks the usual U.S. EPA definition of "immediate analysis", which is 15 minutes, but the PAA concentration has already dropped significantly by then.

4.2 Online Instrumentation

Online instruments are useful tools for establishing disinfection process trends and when real-time dose control is desired. The quality of data gained from online analyzers is directly related to how they are installed and applied at a specific site. Variables such as potential to support biological growth, size and configuration of the sampling system, flowrates through

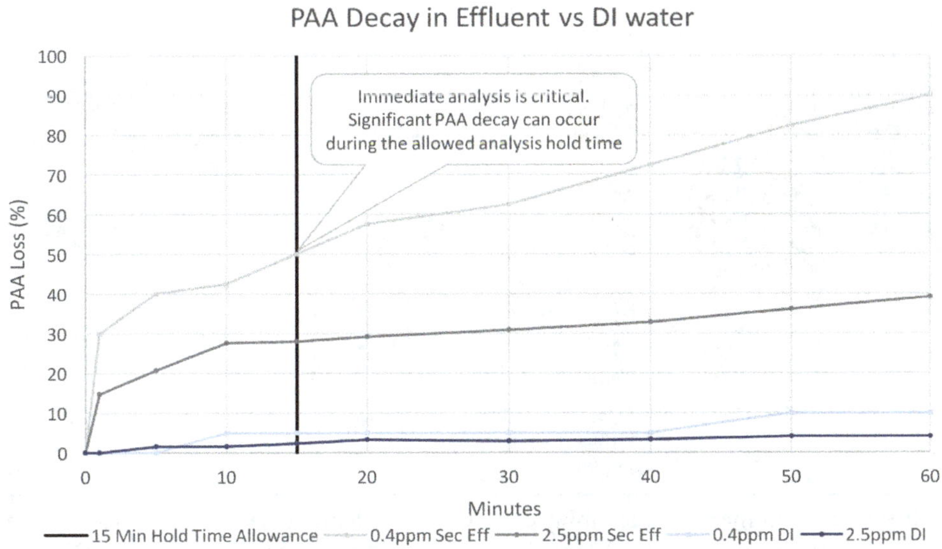

FIGURE 4.4 Decay of two initial concentrations of PAA in wastewater secondary effluent and in deionized water (courtesy of Elisabeth Werth, Metro Wastewater Reclamation District, Colorado).

the sampling system and/or the process flow, analyte concentrations, staffing expertise, and organizational time devoted to instrument maintenance should all be considered when designing for online analyzers.

Biological growth is the largest obstacle to successfully implementing online analyzers in PAA-disinfected streams. For analyzers that require sampling systems with controlled flows and pressures, growth can rapidly accumulate in the sample piping and creates a significant PAA demand between the sampling location in the process stream and the analyzer (Figure 4.5). Figure 4.5 shows field data illustrating PAA demand from growth in sampling tubing. Peracetic acid demand is evident from the drop in concentration at two sampling sites (labeled "north effluent" and "south effluent") detected in the process water compared to the sampling point (halfway through sampling tubing) and the analyzer value.

Certain features of a sampling system are particularly susceptible to biological growth, such as elbows, valves, filters, and sample lines with low velocities. A sampling system designed to be used in PAA-disinfected streams should minimize overall length to reduce sample residence time, eliminate dead or low-flow space, be equipped with cleanouts that provide access

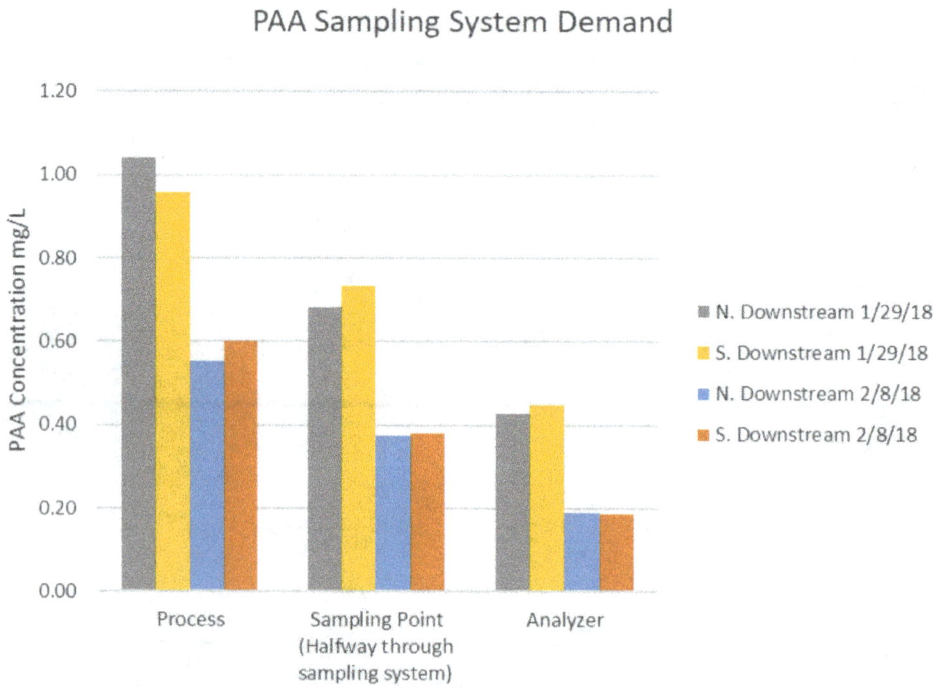

FIGURE 4.5 Online analyzer tubing can exert PAA demand, resulting in inaccurately low PAA readings (courtesy of Ian Myers, Metro Wastewater Reclamation District, Colorado).

to the entire system, use disposable tubing wherever possible, and ideally incorporate a "chemical clean in place" functionality.

Extractive PAA analyzers that use the DPD method are attractive because of their stable calibration and low detection limit; additionally, they do not need to equilibrate after maintenance. Figure 4.6 shows good agreement between samples analyzed using a handheld PAA test kit and an extractive online analyzer. Data quality and consistency from extractive analyzers is, therefore, dependent on the quality of the sample when it reaches the analyzer and the frequency of maintenance performed on the analyzer. More frequent maintenance (daily) can be expected when using a DPD analyzer with small-diameter pipe and fittings. Instruments specifically designed for wastewater applications that use larger-diameter tubing and have self-cleaning capabilities are advised over retrofitting a drinking water instrument.

In situ amperometric probes can be used to eliminate the operations and maintenance complications associated with a sampling system. Amperometric probes are best applied to a process that has PAA concentrations greater than 0.5 mg/L. Although the probes can operate at lower concentrations, they may desensitize if exposed to low or no PAA residuals for prolonged periods of time. In situ probes also require a relatively steady flowrate (0.09 to 0.5 m/s [0.3 to 1.5 ft/s]) across the instrument, which can be difficult to achieve and ensure in retrofitted contact basins.

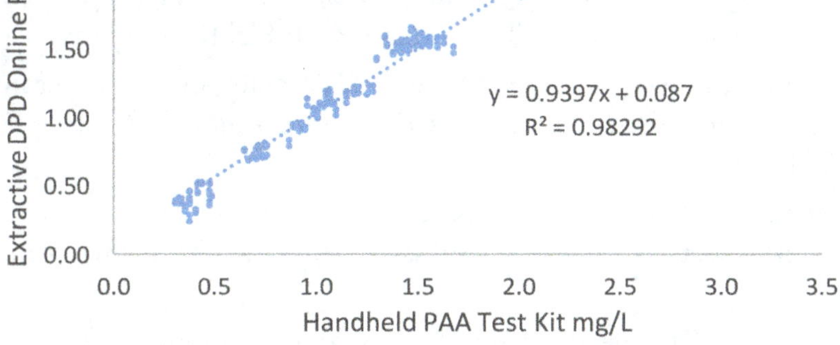

FIGURE 4.6 Example of comparison between a handheld PAA test kit and an extractive online DPD-based PAA analyzer (courtesy of Ian Myers, Metro Wastewater Reclamation District, Colorado).

5.0 REFERENCES

American Public Health Association. (2012). *Standard methods for the examination of water and wastewater* (22nd ed.) Rice, E.W., Baird, R. B., Eaton A. D., & Clesceri, L. S. (Eds.) Washington, DC: American Public Health Association American Water Works Association, & Water Environment Federation.

Analytical Technology, Inc. *Model Q46/85 Peracetic Acid Measurement System Operation and Maintenance Manual*; Rev F (8/17). Collegeville, PA.

Binder, W., & Menger, F. (2000). Assay of peracid in the presence of excess hydrogen peroxide. *Analytical Letters,* 3, 479–488.

Cavallini, G. S., de Campos, S. X., de Souza, J. B., & de Sousa Vidal, M. (2013). Comparison of methodologies for determination of residual peracetic acid in wastewater disinfection. *International Journal of Environmental Analytical Chemistry, 93*(8), 906–918.

CHEMetrics, Inc. (2020) Technical Bulletin, *Hydrogen Peroxide Analysis in the Presence of Peracetic Acid and Copper Using CHEMetrics® Hydrogen Peroxide Test Kits, Cat. Nos. K-5543 and K-5510*, Version 3 Jan. 2020. Retrieved from https://chemetrics.b-cdn.net/uploads/2020/02/peroxide_mod_procedure_PAA_Copper.pdf (accessed March 2020).

Enviro Tech Chemical Services, Inc. (2003). Determination of hydrogen peroxide and peracetic acid in solutions. Modesto, CA.

Graham, M. S., Lomax, G., & Glabra, S. (2012). *Peracetic acid (PAA) versus chlorination/de-chlorination – A disinfection comparison.* Proceedings of the 85th Annual Water Environment Federation Technical Exhibition and Conference; New Orleans, Louisiana, United States, Sept 29–Oct 3. Alexandria, VA: Water Environment Federation. pp. 4672–4684.

Harp, D. L. (1995). *Current technology of chlorine analysis for water and wastewater* (Booklet No. 17). Loveland, CO: Hach Company.

Luukkonen, T., & Pehkonen, S. O. (2017). Peracids in water treatment: A critical review. *Critical Reviews in Environmental Science and Technology*, 1, 1–39

Luukkonen, T., Teeriniemi, J., Prokkola, H., Rämö, J., & Lassi, U. (2014). Chemical aspects of peracetic acid based wastewater disinfection. *Water SA, 40*(1).

Palin, A. T. (1957). The determination of free and combined chlorine in water by the use of diethyl-p-phenylene diamine. *Journal of the American Water Works Association,* 49, 873.

Palin, A. T. (1986). Current DPD methods for disinfectant residual measurement. *Journal of the Institution of Water Engineers and Scientists, 40*(6), 501.

Parker, G. A. (1978). Methods of determination hydrogen peroxide. In G. A. Boltz, D. F., Howell, & J. A. (Eds.), *Colorimetric determination of nonmetals* (2nd ed., Vol. 8, p. 304). New York, NY: John Wiley & Sons.

PeroxyChem. (2014). Old City, New Ideas: Peracetic Acid in Wastewater Disinfection at St. Augustine. *Case Study* Document 04-01-WWT-14.

PeroxyChem. (2015). The decomposition kinetics of peracetic acid and hydrogen peroxide in municipal wastewaters. *Disinfection Forum* No 10, October.

ProMinent Dosiertechnik GmbH *DULCOTEST® PAA*; Part No. 986961; BA DT 097 10/08; 69123 Heidelberg Germany.

Sode, F. (2014). Simultaneous determination of peracetic acid and acectic acid by titration with NaOH. *Analytical Methods*, 6, 2406–2409.

Sully, B. D., & Williams, P. L. (1962). The analysis of solutions of per-acids and hydrogen peroxide. *Analyst*, 87, 653–657.

Wilson, J. (2014). Peracetic acid as an alternative disinfectant. *Clear Waters*, Winter 2014; Syracuse NY: New York Water Environment Association.

6.0 SUGGESTED READING

U.S. Environmental Protection Agency. (1983). *Methods for chemical analysis for water and wastes* (EPA-600/4-79/020). Washington, DC: U.S. Environmental Protection Agency.

U.S. Environmental Protection Agency. (2002). *Method 1603: Escherichia coli (E. coli) in water by membrane filtration using modified membrane-thermotolerant escherichia coli agar (modified mTEC)*. Washington, DC: U.S. Environmental Protection Agency, Office of Water.

U.S. Environmental Protection Agency. (2015). *Review of coliphages as possible indicators of fecal contamination for ambient water quality* (EPA-820-R-15-098). Washington, DC: U.S. Environmental Protection Agency, Office of Water.

5

Effluent Toxicity and Calculation of Allowable Residuals

Philip Block, Daniela Castañeda, Kati Bell, & Vince Chiu

1.0 THE ENVIRONMENTAL FATE OF PERACETIC ACID

The aquatic fate of peracetic acid (PAA) is almost exclusively determined by degradation in the water phase and is driven by several mechanisms, including hydrolysis, auto-decomposition, and reaction with organics and reduced metals. The addition of chemical oxygen demand, biochemical oxygen demand, or total suspended solids increases the degradation rate of PAA, thereby reducing the longevity of PAA in the environment and its impact on aquatic life. Thus, PAA does not persist long in the environment and decomposes to relatively benign components (water, acetic acid, and oxygen).

The low octanol-water partition coefficient (K_{ow}) of PAA and hydrogen peroxide, 0.3 and 0.4, respectively, indicate bioaccumulation in aquatic organisms is not likely. Additionally, the degradation product acetic acid will not bioaccumulate (K_{ow} = 0.68). The soil–sediment adsorption coefficients are also low. The U.S. Environmental Protection Agency (U.S. EPA) concluded that there is no potential for bioaccumulation or bioconcentration of PAA (U.S. EPA, 2010). As a result, the environmental impact from PAA will be minimal and short-lived.

1.1 Available Literature on Organism Toxicity

1.1.1 Peracetic Acid

In 2001, the European Centre for Ecotoxicology and Toxicology of Chemicals, Joint Assessment of Commodity Chemicals (JACC) program, published JACC No. 40, *Peracetic acid and its equilibrium solutions* (European Centre for Ecotoxicology and Toxicology of Chemicals, 2001). This report presents a comprehensive review of PAA, its distribution in the environment, and its impact on aquatic life. The report shows the effect of various PAA formulations on standard aquatic species conducted according to guideline laboratory study protocols. In addition, the Organization for Economic Cooperation and Development (OECD), in cooperation with the U.S. EPA's High Production Volume (HPV) program, published a Screening Information Data Set (SIDS) for PAA (OECD, 2008).

From laboratory data, a relationship is observed between the size of the organism and its sensitivity to PAA. The smaller the test organism, the greater the sensitivity as compared to larger test organisms. It is suggested that small organisms are more sensitive to PAA because their body surface to body weight ratio is high, which enables a higher uptake of PAA per gram of body weight. Additionally, longer exposure times lead to more PAA toxicity to the organism. Peracetic acid toxicity to marine-estuarine organisms is also found to be lower compared to freshwater species because of its shorter half-life (<1 hour) in the saltwater environment (Koivunen & Heinonen-Tanski, 2005).

U.S. EPA has classified the aquatic toxicity of PAA as toxic to algae *(Selenastrum capricornutum)* and aquatic invertebrates *(Daphnia magna)*, and slightly to moderately toxic to fish based on laboratory test results. However, PAA is less impactful to aquatic organisms than chlorine, which is classified as highly toxic to all aquatic lifeforms. Table 5.1 presents a summary of acute toxicity data from several sources compiled for both PAA and sodium hypochlorite; the data summarized have been developed using standardized protocols for establishing the following values:

- EC_{50}—the dose causing incapacitation to 50% of the population tested, determined by running several doses and estimating the EC_{50} statistically;

TABLE 5.1 Acute toxicity of PAA and sodium hypochlorite (reprinted with permission of PeroxyChem, LLC).

Acute Toxicity			
96-hour LC_{50} Data			
Species		Peracetic acid, 15%[a,b,c]	Sodium hypochlorite, 12%[d,e,f,g]
Oncorhynchus mykiss, Rainbow trout		0.53–1.6 mg/L	0.03–0.22 mg/L
Lepomis macrochirus, Bluegill sunfish		1.1–3.3 mg/L	0.44–0.88 mg/L
Salvelinus fontinalis, Brook trout		Not available	0.1–0.18 mg/L
Lepomis cyanellus, Green sunfish		Not available	0.71–0.82 mg/L
Brachydanio rerio, Zebrafish		1.0 mg/L	N/A
EC_{50} Data			
Daphnia magna	48 hour	0.73–3.3 mg/L	0.033–0.048 mg/L
Selenastrum capriconutum	120 hour	0.18 mg/L	Toxic to aquatic organisms
Mytilus edulis embryo	48 hour	0.27 mg/L	N/A
Algae, various species	72–96 hour	0.7–16 mg/L	N/A

[a]European Centre for Ecotoxicology and Toxicology of Chemicals, 2001
[b]OECD, 2008
[c]Health Canada, 2013
[d]OECD, 2004
[e]Pesticide Action Network, 2012
[f]Gurtler Industries, Inc., 2015
[g]U.S. EPA, 2010

- EC_0—the dose causing no incapacitation to the population tested;
- LC_{50}—the dose causing lethality to 50% of the population tested, determined by running several doses and estimating the LC_{50} statistically; and
- LC_0—the dose causing no lethality to the population tested.

The acute data that have been summarized here show that PAA is less toxic to organisms, relative to chlorine, as indicated by the higher concentrations that are required to elicit a particular toxic response. Chronic toxicity data (presented in Table 5.2) are compiled for additional endpoints as follows:

- IC_{50}—the concentration of a substance that inhibits specific biological function in 50% of the test population;
- NOEC—no observable effect concentration; or the highest dose tested that does not cause any toxic effects to the organisms tested; and
- LOEC—lowest observable effect concentration; this is the lowest dose tested that is statistically significant over a specified exposure time.

The Screening Information Data Set (SIDS) prepared by The Organization for Economic Co-operation and Development (OECD, 2008) concluded the following:

TABLE 5.2 Chronic toxicity of PAA (reprinted with permission of PeroxyChem, LLC).

96-hour LC_{50} data				
Species	**Test design and endpoint**	**LC_{50} or IC_{50}**	**NOEC**	**LOEC**
P. promelas, Fathead minnow	Flow-through Survival	1.16 mg/L	0.89 mg/L	1.62 mg/L
P. promelas, Fathead minnow	Flow-through Growth	1.19 mg/L	0.89 mg/L	1.62 mg/L
C. dubia, Water flea	Flow-through Survival	0.86 mg/L	0.68 mg/L	1.45 mg/L
C. dubia, Water flea	Flow-through Growth	1.03 mg/L	0.68 mg/L	1.45 mg/L
S. capriconutum, Green algae	Static renewal Growth	0.44 mg/L	0.19 mg/L	0.38 mg/L
A. flosaquae, Blue-green algae	Static renewal Growth	0.49 mg/L	0.20 mg/L	0.40 mg/L

- The 96-hour LC_{50} values for fish ranged between 0.9 and 3.3 mg/L in most freshwater species; the 48-hour EC_{50} for *D. magna* ranged between 0.5 and 1.0 mg/L.
- Based on the representative standard toxicity tests, the lowest 72-hr NOEC of 0.084 mg/L was found for a freshwater, green algae called *Pseudokirchneriella subcapitata* (formerly known as *Selenastrum capricornutum*).
- The lowest EC_{50} value of 0.18 mg/L was found during a 120-hr growth inhibition test with *P. subcapitata*.

Work by Chhetri et al. (2017) determined an EC_{50} for *P. subcapitata* of 1.38 mg PAA/L (for comparative purposes, they found EC_{50}s of 0.34 and 0.16 for performic acid and for chlorine dioxide, respectively).

Peracetic acid has been studied extensively for use in aquaculture and fish farming (Hushangi & Shekarabi, 2018; Liu, Behrens, Pedersen, Straus, & Menelt, 2016; Liu, Straus, Pedersen, & Meinelt, 2018; Liu, 2018; Pedersen, Meinelt, & Straus, 2015; Straus, 2018; Straus, Meinelt, Farmer, & Beck, 2012; Straus, Meinelt, Liu, & Pedersen, 2017) for the treatment of freshwater fish pathogens, such as gill mites. As an example, Straus et al. (2017) investigated PAA toxicity on 12 species of fingerling fish via 24-hr static toxicity bioassays. The LC_{50} values ranged from 2.8 to 9.3 mg/L PAA, and the NOEC values ranged from 1.9 to 5.8 mg/L PAA. Black fathead minnows were the most sensitive, and blue tilapia were the most tolerant to PAA exposure; the LC_{50} values of other species tested were within the range of 4.1 to 6.2 mg/L PAA.

1.1.2 Hydrogen Peroxide

Because PAA exists in equilibrium with hydrogen peroxide, and it is not possible to generate a pure PAA compound, aquatic toxicity testing of PAA incorporates the influence of hydrogen peroxide in the formulation. Upon dilution into the receiving stream, the hydrogen peroxide component may spatially separate out from the PAA due to the difference in decomposition rates (the peroxide may survive downstream longer than the PAA because of slower decomposition kinetics). However, the aquatic toxicity effects of the resultant hydrogen peroxide cannot exceed the effects of the combined PAA/hydrogen peroxide system that is typically measured in PAA toxicity studies. Yet, there is interest in the impact of the hydrogen peroxide on aquatic toxicity as delivered from a PAA solution.

Hydrogen peroxide has been thoroughly studied for acute toxicity to freshwater fish and aquatic invertebrates (European Centre for Ecotoxicology and Toxicology of Chemicals, 1993; European Chemicals Branch, 2003).

Studies have been conducted from 24 to 96 hours in aqueous media. None of the existing studies report results after 168 hours (7 days) of exposure because of the rapidity of hydrogen peroxide degradation, whose half-life is 7.8 hours in unfiltered lake water (Cooper & Lean, 1989). Table 5.3 summarizes the aquatic toxicity studies in the literature for hydrogen peroxide.

TABLE 5.3 Hydrogen peroxide aquatic toxicity for a variety of species.

Species	Effect		Concentration (mg/L)	Reference
Freshwater Crustacea				
Gammarus	LC_{50}	96 hr	4.42	Kay, Quimby, & Ouzis, 1982
Daphnia pulex	LC_{50}	48 hr	2.4	Shurtleff, 1989
Daphnia pulex	NOEC		1.0	Shurtleff, 1989
Daphnia pulex	Immobilization	5 min	4.2	Gannon & Gannon, 1975
Daphnia magna	EC_{50} immobilization	4 hr	7.7	Bringmann & Kuehn, 1982
	EC_0		3.8	
P. clarkia (crayfish)	LC_0		64.6	Kay et al., 1982
Freshwater Fish				
Gambusia affinis (Mosquito fish)	Unharmed, field study		2.38–9.86	Kay et al., 1982
Ictalurus punctatus (Channel catfish)	LC_{50}	96 hr	37.4	Kay et al., 1982
Lebistes reticulatus (Guppy)	Unharmed		34	Quimby, 1961
Leuciscus idus melanotus (Golden orfe)	LC_{50}	24 hr	35	Degussa, 1977
Pimephales promelas (Fathead minnow)	LC_{50}	96 hr	16.4	Shurtleff, 1989
	NOEC		5	
Ptychocheilus oregonensis (Squawfish)	LC_0	24 hr	10	McKee & Wolf, 1963
Salmo gairdneri (Rainbow trout fingerling)	LC_{50}	48 hr	>40	Southgate, 1950

Chhetri, Baun, and Andersen (2017) measured a hydrogen peroxide EC_{50} for the algal species *P. subcapitata* of 2.90 mg/L (compared to 1.38 for PAA). In general, the aquatic toxicity of hydrogen peroxide does not become prevalent until concentrations that are 10 to 100 times greater than that of PAA. Liu et al. (2015) investigated 24-hour toxicity of PAA formulations with varying amounts of hydrogen peroxide on *Daphnia magna*. They concluded that the toxicity of PAA solutions is attributable to an additive effect of PAA and hydrogen peroxide, with increasing concentration of peroxide in the formulation having greater toxicity for the same PAA dose. However, in contrast to this outcome, the OECD SIDS concluded the following: "In general, the aquatic test with fish, invertebrates and algae were reproducible if concentrations were expressed as peracetic acid irrespective of the concentrations of hydrogen peroxide and peracetic acid. Thus, the peracetic acid concentration alone may explain the toxicity of PAA formulations" (OECD, 2008).

2.0 PERMISSIBLE PERACETIC ACID RESIDUALS TO THE RECEIVING STREAM

2.1 Generation of the Federal Non-Site-Specific Permissible Peracetic Acid Residual Limit

PAA represents one of the more recent technologies that has undergone review by U.S. EPA and the Canadian Pest Management Regulatory Agency (PMRA) in the process of obtaining regulatory approval of PAA as a wastewater disinfectant. The process by which U.S. EPA determined the allowable PAA residual to enter the environment post-outfall from the treatment facility and the steps to obtain PAA product label registration involved several different divisions within the agency and took several years to complete. The review process and results are described herein.

2.1.1 U.S. EPA Review of Peracetic Acid

In the United States, biocides, such as PAA, are regulated under the Federal Insecticide, Fungicide and Rodenticide Act (FIFRA, 7 U.S.C. Section 136), requiring registration of the biocide with the U.S. EPA. In 2002, the U.S. EPA Office of Water *deferred* regulation of wastewater disinfection by PAA to the U.S. EPA Office of Chemical Safety and Pollution Prevention/Office of Pesticide Programs, Antimicrobials Division. The Antimicrobials Division prepared data requirements for PAA to treat municipal wastewater effluent. The Office of Water also participated in the requirements-setting meeting held in October 2002; the requirements were issued in November of that year.

The acute toxicity endpoint, as determined by good laboratory practice study, for the most sensitive aquatic species to PAA, *Selenastrum capricornutum* (green algae), is an EC_{50} of 0.18 ppm. If the PAA effluent discharge is less than the expected environmental concentration (EEC) of 0.2 ppm (predetermined by federal, state, and local law before water is discharged into a waterway), the potential risk to even the most sensitive aquatic organisms will be minimal, according to the U.S. EPA's level of concern (LOC) scheme in estimating risk for aquatic species:

- Acute Risk: potential for acute risk for all non-target organisms, >0.5
- Acute Restricted Use: potential for acute risk for all non-target organisms, but may be mitigated through restricted use classification, >0.1
- Acute Endangered Species: endangered species may be adversely affected by use, >0.05
- Chronic Risk: potential for chronic risk may warrant regulatory action, >1

The 0.2-ppm level is based on a worst-case, Tier I screening-level risk assessment. It is not site-specific and did not consider subsequent dilution and degradation that would rapidly occur. In practice, U.S. EPA accepted that a maximum 1-ppm PAA discharge from a wastewater facility will degrade easily and rapidly to hydrogen peroxide, acetic acid, and water, and that the concentration at 1 ppm would quickly be reduced to below 0.2 ppm. In part, this conclusion is supported by the analysis found in the following sections.

2.1.2 Influence of Peracetic Acid Decomposition, Dispersion, and Dilution

To investigate the impact of PAA decomposition, dispersion, and dilution upon the downstream PAA concentration in receiving bodies after the facility outfall, a computational modeling study (MWH Americas, 2014) was performed for three different receiving bodies (streams, lakes, and ocean). For this study, determination of the mixing zone was set by assuming a very conservative water quality standard with a PAA concentration of 0.065 mg/L, starting with an outfall concentration of 1.0 mg/L. The study focused on near-field mixing, which is influenced by the physical characteristics of the effluent and receiving body, such as flowrate, buoyancy, velocity, and similar parameters. Fair-field mixing, induced by wind currents, eddies, and such, were not accounted for in this model.

Simulations for the study were performed by the Cornell Mixing Zone Expert System (CORMIX) (Doneker & Jirka, 1990; Frick, Roberts, Davis,

Keyes, Baumgartner, & George, 2003). CORMIX is a U.S. EPA-supported simulation tool for environmental assessment of mixing zones and dilution from point sources. The model evaluated the PAA discharge into the receiving body via a submerged pipe. A PAA first-order decay constant of 10.08 day^{-1} (Rossi, Antonelli, Mezzanotte, & Nurizzo, 2007) was chosen. Ambient flow velocities for the evaluation varied between 0.2 and 1.5 m/s for the stream, 0.5 to 1 m/s for the lake, and 0.5 to 0.75 m/s for the ocean; the facility discharge rate was set at either 18.9 or 94.6 ML/d. The outfall varied from 0.5 to 0.8 m in diameter, the terminus of which was submerged and positioned at the bottom of the waterbody.

The results of the study demonstrated that the PAA discharged into the receiving waterbody dissipates within a relatively short distance downstream of the discharge point. For the "stream receiving body" simulation, the maximum distance downstream to reach the target mixing zone concentration of 0.065 mg/L with a PAA discharge concentration of 1 mg/L ranged from 2.1 m (low discharge rate into high ambient velocity) to 78.6 m (high discharge rate into low ambient velocity), the latter representing the worst-case scenario. The lake and ocean receiving body simulations demonstrated that the minimum distances to reach the target PAA concentration were much lower than that for the stream model. The study demonstrated that the PAA residual time in the receiving body will be short and the resultant impact on the environment minimal.

2.2 Determination of Residuals Dilution Factors and Discharge Concentrations Absent In-Stream Criteria

Non-site-specific, allowable residuals do not account for the dilution effect of the receiving stream in reducing the PAA concentration to below acceptable toxic levels. A computer simulation such as that described in the last section demonstrates that diffusion and dilution, along with decomposition, can significantly reduce the effective range of PAA in the environment. The addition of dilution factors in modeling can account for this effect and can be used in site-specific determinations of allowable PAA residuals.

2.2.1 Dilution Factors as a Function of Effluent Discharge and Receiving Water Flowrates

Dilution factors for wastewater treatment facilities (WRRFs) can be calculated using the facility flowrates and the 7Q10 flowrates in the receiving waterbodies. The 7Q10 flowrate is the lowest average flow for a receiving stream for a seven-day period over the last 10 years, and it is commonly used for engineering design work (Thomann & Mueller, 1987) as a worst-case scenario of minimal stream flow. Dilution factors (DF) are described by the following equation:

$$DF = \frac{\text{facility effluent discharge} + \text{receiving stream 7Q10}}{\text{facility effluent discharge}} \tag{5.1}$$

U.S. EPA provides screening-level estimates of the concentration of chemicals from consumer products released to the air, surface water, and landfills via its E-FAST program (U.S. EPA, n.d.). A 2009 review of E-FAST showed the dataset contained 27 249 records, of which 19 169 are for Standard Industrial Code 4952, the designation for a WRRF. The dataset was not complete, however, so there were only 9026 records with both facility discharge rate data and 7Q10 flowrates and 9242 records with both facility discharge rate and mean flowrate data. Table 5.4 shows these data, along with the resulting dilution factor.

TABLE 5.4 Dilution factors for identified WRRFs (reprinted with permission of PeroxyChem, LLC).

	Dilution factor	
Percentile	**7Q10 flowrate**	**Mean flowrate**
Minimum	1.000 029	1.033
5th	1.082	13.7
10th	1.31	33.0
20th	2.5	91.6
25th	3.7	133
30th	5.6	186
40th	12	344
50th	25	611
60th	55	1161
70th	125	2283
80th	347	5625
90th	1592	20 717
95th	6440	68 048
Maximum	123 565 116	572 000 000

2.2.2 *Peracetic Acid Discharge Limits Based on Dilution Factors*

The concentration of PAA in the receiving stream can be calculated from the dilution factor (DF) by Equation 5.2, as follows:

$$\text{PAA Concentration in the receiving water} = \text{PAA Concentration in Effluent/DF} \quad (5.2)$$

As described earlier, the PAA concentration 0.18 mg/L is protective of the most sensitive aquatic species, algae. If a risk quotient factor (RQF) of 2 is applied, then a very conservative PAA concentration target in the receiving stream is 0.09 mg/L. Applying Equation 5.2 to this target PAA concentration provides for the following dilution-factor-based, site-specific PAA effluent limits shown in Table 5.5.

2.3 Recommended Residuals by Regulatory Agencies

As previously discussed, the U.S. EPA's approval of PAA wastewater disinfection with an effluent discharge of 1.0 mg/L is based on an assessment of toxicological risk and PAA longevity in the environment. At the time of

TABLE 5.5 Dilution-factor-based PAA discharge limits based on a target concentration of 0.09 mg/L (reprinted with permission of PeroxyChem, LLC).

	PAA effluent discharge limit (mg/L)	
Percentile	**7Q10 flowrate**	**Mean flowrate**
Minimum	0.090	0.93
5th	0.97	1.24
10th	0.118	2.97
20th	0.229	8.25
25th	0.333	12.0
30th	0.501	16.8
40th	1.06	21.0
50th	2.26	55.0
60th	4.94	104
70th	11.3	205
80th	31.2	506
90th	143	1865
95th	580	6124

this publication, U.S. EPA had also approved at least one commercial PAA product to have dilution-factor-based, site-specific maximum PAA effluent concentrations based on the following calculation:

$$\text{PAA maximum effluent concentration} = 0.9 \times \text{DF; for DF} \geq 12$$
$$= 1.0 \text{ for DF} < 12 \text{ or if 7Q10 flowrate not known.}$$

Each U.S. state may regulate tighter restrictions on the allowable maximum PAA discharge concentration, and, as a result, PAA approval for use is performed on a state-by-state basis. Currently, and subject to change in the coming years, many states have not yet set an allowable PAA residual. Several states have adopted the U.S. EPA non-site-specific PAA maximum effluent concentration of 1.0 mg/L, including Alabama, Florida, Missouri, New Jersey, Oklahoma, and Oregon. Several states have set lower permissible effluent concentrations, such as Colorado, Iowa, Ohio, and Tennessee, although Tennessee has taken a more site-to-site approach and will allow greater than 1.0 mg/L effluent concentrations if the receiving stream has a high dilution factor. Kentucky is currently approving permit modifications on a facility-by-facility basis.

The PMRA in Canada granted registration approval for PAA as a wastewater disinfectant for use in that country. As with the United States, each province then has regulatory authority to set permissible PAA effluent concentrations. British Columbia has approved a PAA discharge limit of 1.0 mg/L, but at the time of this publication, no other Canadian province has allowed a permit modification for use of PAA. European countries have taken the lead on approving the use of PAA as the use of hypochlorite has come under pressure because of regulations starting in 1999 that reduced the allowable concentrations of disinfection byproducts in the environment.

3.0 WHOLE EFFLUENT TOXICITY TESTING

Whole effluent toxicity (WET) refers to the aggregate toxic effect to aquatic organisms from all pollutants contained in a facility's wastewater effluent. Whole effluent toxicity tests measure wastewater's effects on specific test organisms' ability to survive, grow, and reproduce. U.S. EPA's WET methods are listed in 40 C.F.R. 136.3. The WET test procedures can be found in U.S. EPA's *Methods for Measuring the Acute Toxicity Effects of Effluents and Receiving Waters to Freshwater and Marine Organisms (*U.S. EPA, 2002), and the U.S. EPA publication *Method Guidance and Recommendations for Whole Effluent Toxicity (WET) Testing* (U.S. EPA, 2000) provides guidance on performing such tests. Table 5.6 lists the U.S. EPA methods for WET (40 C.F.R. 136.3).

In general, WET testing requires generating a dose response between the compound in question, in this case PAA, and the effect being monitored for, such as lethality. However, the very nature of PAA and its lack of stability in effluents and receiving streams, while a benefit in terms of significantly truncating its long-term impact on the environment, makes WET results

TABLE 5.6 U.S. EPA test methods for whole effluent toxicity testing.

Parameter	Method	U.S. EPA no.
Acute Toxicity fresh water organisms LC_{50}, percent effluent	*Ceriodaphnia dubia* acute	2002.0
	Daphnia puplex and *Daphnia magna* acute	2021.0
	Fathead minnow, *Pimephales promelas,* and Bannerfin shiner, *Cyprinella leedsi,* acute	2000.0
	Rainbow trout, *Oncorhynchus mykiss,* and brook trout, *Salvelinus fontinalis,* acute	2019.0
Acute Toxicity estuarine and marine organisms LC_{50}, percent effluent	Mysid, *Mysidopsis bahia,* acute	2007.0
	Sheepshead minnow, *Cyprinodon variegatus,* acute	2004.0
	Silverside, *Menidia beryllina, Menidia,* and *Menidia peninsulae,* acute	2006.0
Chronic Toxicity fresh water organisms NOEC or IC_{25}, percent effluent	Fathead minnow, *Pimephales promelas,* larval survival and growth	1000.0
	Fathead minnow, *Pimephales promelas,* embryo-larval survival and teratogenicity	1001.0
	Daphnia, *Ceriodaphnia dubia,* survival and reproduction	1002.0
	Green alga, *Selenastrum capricornutum,* growth	1003.0
Chronic Toxicity estuarine and marine organisms NOEC or IC_{25}, percent effluent	Sheepshead minnow, *Cyprinodon variegatus,* larval survival and growth	1004.0
	Sheepshead minnow, *Cyprinodon variegatus,* embryo-larval survival and teratogenicity	1005.0
	Inland silverside, *Menidia beryllina,* larval survival and growth	1006.0
	Mysid, *Mysidopsis bahia,* survival, growth, and fecundity	1007.0
	Sea urchin, *Arbacia unctulate,* fertilization	1008.0

difficult to interpret because a PAA residual cannot be maintained. Hence, PAA's lack of long-term impact on the receiving stream flora and fauna makes it difficult to measure that very impact in a standard WET procedure.

3.1 Chronic and Acute Whole Effluent Toxicity

Toxicity studies with PAA-disinfected wastewaters have been performed by several WRRFs as part of their data submittal packages to their state regulatory agencies for their permit modification request. Outcomes of these tests are exemplified in a recent study (Water Research Foundation, 2018), during which PAA disinfection pilot reactor trials were performed. Whole effluent toxicity testing was performed on two species: *Ceriodaphnia dubia* (water flea) and *Pimephales promelas* (fathead minnow).

For *C. dubia*, U.S. EPA Method 1002 was used, which is a seven-day, chronic survival and reproduction test. Five concentrations were tested: 12%, 17%, 23%, 30%, and 40% of the effluent. For *P. promelas*, U.S. EPA Method 1000 was used, which is a seven-day, chronic survival and growth test. Five concentrations were tested: 12%, 17%, 23%, 30%, and 40% of the effluent. For both, data were statistically analyzed ($p = 0.05$, per U.S. EPA methodology) for LOEC and NOEC. For all four treatment facilities in this study, initial PAA dose concentrations ranged from 1.5 to 3.5 mg PAA/L, depending on the facility, and effluent samples were collected after 30 minutes of contact time. No observable effect on either survival or reproduction/growth for both *P. promelas* and *C. dubia* was observed in any sample, the result being that no LOECs were able to be calculated. In general, these results are typical, and reflect the short lifetime of PAA in the effluent.

3.2 Static Renewal and Continuous Dosing Studies

To address the issue of the rapid PAA decay in wastewater influent, a seven-day renewal, chronic toxicity study was performed (PeroxyChem, 2015) using a 15% PAA and 23% hydrogen peroxide formulation, under the methodology proposed by the Tennessee Department of the Environment and Conservation. The PAA was renewed daily at dose concentrations of 0.075, 0.15, 0.30, 0.60, and 1.2 mg PAA/L. The results demonstrated that the IC_{25} for *C. dubia* reproduction was 0.38 ppm of PAA, and for *P. promelas* biomass the IC_{25} was 0.75 ppm. The NOEC determined for fish survival and biomass was 0.6 ppm, and for *C. dubia* survival it was 0.3 ppm.

A similar study conducted a side-by-side trial of municipal wastewater with one contact channel treated with PAA and a second channel by sodium hypochlorite with subsequent quenching of the chlorine with sodium bisulfite (Keogh & Tranh, 2011). One treatment train was dosed with 8 mg/L of chlorine, followed by dechlorination, while the second train was dosed with 1.5 mg/L PAA. These disinfectant doses were chosen to achieve equivalent microbial reductions needed to meet the facility's permitted effluent requirements. A seven-day renewal study was performed with effluents from both treatment trains at dilutions of 6.25%, 12.5%, 25%, 50%, and 100% of the effluent. Two species were selected for assessment: *Mysidopsis bahia* (mysid shrimp) and *Menidia beryllina* (tidewater silverside). Table 5.7 shows the survival rate results from the study. Growth rate results (not shown in the table) were equivalent for PAA and hypochlorite/dechlorination.

These results demonstrate that PAA has a similar, if not slightly better, toxicity profile in this wastewater as compared to the wastewater that was chlorinated with subsequent addition of a quenching agent.

4.0 QUENCHING CONSIDERATIONS

If the PAA residual exceeds either the federally allowable maximum per the product label or the state-approved limit via permit, excess PAA in the waste stream can be quenched in a manner analogous to hypochlorite through the addition of a reducing agent. Common reducing agents that can be used to quench PAA include sodium bisulfite (SBS), sodium metabisulfite (SMBS), and calcium thiosulfate (CTS).

The stoichiometric amount of the quenching agent to reduce the residual PAA is as follows:

For sodium bisulfite ($NaHSO_3$), the mole ratio of SBS to PAA is 1:1.

$$NaHSO_3 + CH_3COOOH \rightarrow CH_3COOH + NaHSO_4 \qquad (5.3)$$

TABLE 5.7 Seven-day renewal results for survival rate for PAA and chlorine followed by dechlorination (reprinted with permission of PeroxyChem, LLC).

% survival	*M. bahia*		*M. beryllina*	
% effluent	PAA	Chlorine/dechlor	PAA	Chlorine/dechlor
Control	92.5	97.5	100	100
6.25	90	85	100	97.5
12.5	95	97.5	100	97.5
25	95	90	100	100
50	90	77.5	100	100
100	97.5	82.5	97.5	97.5

PAA has a molecular weight of 76.05 g/mol, and SBS has a molecular weight of 104.1 g/mol. To quench 1 ppm of PAA, 1.37 ppm (mg/L) of SBS is needed.

For sodium metabisulfite ($Na_2S_2O_5$), the mole ratio of SMBS to PAA is 1:2.

$$Na_2S_2O_5 + H_2O \rightarrow 2\ NaH_SO_3 \quad (5.4)$$

$$2\ NaHSO_3 + 2\ CH_3OOOH \rightarrow 2\ CH_3COOH + 2\ NaHSO_4 \quad (5.5)$$

The molecular weight of SMBS is 190.1 g/mol. Thus, to quench 1 mg/L of PAA, 1.25 mg/L of SMBS is needed.

For calcium thiosulfate (CaS_2O_3), the mole ratio of CTS to PAA is 1:4.

$$CaS_2O_3 + 4\ CH_3COOOH + H_2O \rightarrow 4\ CH_3COOH + Ca(HSO_4)_2 \quad (5.6)$$

The molecular weight of CTS is 152.1 g/mol. To quench 1 ppm of PAA, 0.5 mg/L of CTS is needed.

The hydrogen peroxide in the formulation may require an additional quenching agent because it will provide an oxidant sink for the reducing agent. The kinetics of reaction between hydrogen peroxide and the quenching agent may be slower than that for PAA and the reductant; thus, estimation of the effect of hydrogen peroxide on reductant demand may be considered conservative. The stoichiometry for hydrogen peroxide is similar to PAA, as shown in Equations 5.3 to 5.6, and the molecular weight of hydrogen peroxide is 18 g/mol. To quench 1 mg/L of hydrogen peroxide requires 5.8 mg SBS/L, 5.3 g SMBS/L, or 2.1 g CTS/L.

5.0 RESIDUAL IMPACTS ON THE RECEIVING STREAM

Use of PAA for wastewater disinfection may deliver PAA, hydrogen peroxide, and acetic acid into the receiving stream. In addition, water and oxygen will be generated within the effluent or receiving stream. The octanol/water partitioning coefficient for PAA is 0.3 (0.4 for hydrogen peroxide) (European Centre for Ecotoxicology and Toxicology of Chemicals, 2001), and thus is not considered to bio-accumulate or partition into soil systems in the environment. Soil column experiments using 2000 mg/L PAA showed that 21.5% of the PAA was recovered 25 mm into the soil, and that only 3% was recovered at a depth of 50 mm (10% of the hydrogen peroxide was recovered at 50 mm) (European Centre for Ecotoxicology and Toxicology of Chemicals, 2001). This indicates a rapid reaction between PAA and soil organic constituents, which prevents significant penetration into the soil subsurface.

Peracetic acid and hydrogen peroxide will oxidize reduced metals, particularly iron, to their oxide forms, which, in turn, is one of the primary drivers of PAA decomposition in the wastewater and receiving stream (Henao, Compagni, Turolla, & Antonelli, 2018). Inorganic compounds containing nitrogen (e.g., nitrate, nitrite, and ammonia) and phosphorus have been shown to have little interaction with PAA with respect to putting a demand on PAA (Henao et al., 2018) and thus are expected to have little interaction in the environment after the facility outfall. In this same study, PAA was shown to react with protein structures, such as casein.

5.1 Biological Oxygen Demand

Because PAA contains acetic acid, it will generate biological oxygen demand (BOD) per the following equation:

$$CH_3COOH + 2\,O_2 \rightarrow 2\,CO_2 + 2\,H_2O \tag{5.7}$$

There are several different formulations of PAA on the market, each with a different ratio of PAA to hydrogen peroxide to acetic acid. The different formulations will generate a different amount of theoretical biological oxygen demand, as shown in Table 5.8.

The actual effect of PAA on receiving stream BOD levels is not straightforward, with indications that, under some circumstances, BOD values are seen to increase with the addition of PAA. Others observed a decrease in BOD and, additionally, no change has been reported. One such study investigated the impact of PAA on BOD and observed that a dosage of 10 mg PAA/L showed no significant change in the BOD of the effluent (Cavallini & de Souza, 2013). Luukkonen, Teeriniemi, Proddola, and Lassi (2014) measured BOD following PAA addition up to 8 mg/L to tertiary treated wastewater in a pilot facility setting in Finland. Measured changes in wastewater

TABLE 5.8 Theoretical BOD generated by various commercial PAA formulations (reprinted with permission of PeroxyChem, LLC).

PAA wt%	Hydrogen peroxide wt%	Acetic acid wt%	Theoretical BOD (mg O_2/L)
15	23	16	1.98
12	18.5	20	2.62
22	5	45	3.02

BOD after PAA addition were significantly lower than the theoretical BOD (and even showed a decrease in BOD at PAA concentrations below 2 ppm).

This echoed work by Baldry, Cavadore, French, Massa, Rodrigues, Schirch, and Threadgold (1995), in which BOD was observed to fall during full-scale PAA trialing. This was possibly attributed to the oxidation of organics by PAA in the textile-constituent wastewater. To the point, PAA may aid in the reduction of BOD because of the oxidation of wastewater organics into smaller constituents. Stampi, de Luca, and Zanetti (2001) reported virtually no change, within the standard error, of BOD_5 values in wastewater from a facility using PAA as a disinfectant (contact time 20 minutes with a PAA dose of 1.5 to 2 ppm) pre- and post-PAA disinfection.

5.2 Dissolved Oxygen

In addition to biological oxidant demand, PAA formulations can also generate dissolved oxygen (DO) via Equations 5.7 and 5.8, as follows:

$$2\ CH_3CO_3H \rightarrow 2\ CH_3CO_2H + O_2 \tag{5.7}$$

$$2\ H_2O_2 \rightarrow 2\ H_2O + O_2 \tag{5.8}$$

Table 5.9 shows the generated DO for PAA solutions of varying formulations. The "apparent BOD" is defined as the theoretical BOD from Table 5.8 minus the generated DO from Equations 5.7 and 5.8.

Cavallini and de Souza (2013) performed laboratory testing of PAA in wastewater obtained from the outfall of a treatment facility in Brazil. Their results demonstrated that DO increased within increasing dosages of PAA at several different temperatures, up to 40 mg PAA/L. Above these PAA concentrations, they observed that the DO levels plateaued, which they attributed to supersaturation of the effluent. The research concluded that "since DO is

TABLE 5.9 Theoretical generated DO and apparent BOD from commercially available PAA formulations (reprinted with permission of PeroxyChem, LLC).

PAA wt%	Hydrogen peroxide wt%	Acetic acid wt%	Generated DO (mg O_2/L)	Apparent BOD (theoretical BOD minus DO)
15	23	16	0.93	1.05
12	18.5	20	0.94	1.68
22	5	45	0.32	2.70

a parameter of great relevance to the quality of bodies of water, the release of effluents with high DO concentrations could be another advantage of PAA application, in addition to deactivation of microorganisms".

5.3 pH

Peracetic acid solutions have a pH less than 1 s.u. However, at the typical dose rates of 0.5 to 3 mg PAA/L wastewater, it is common to see little to no change in the pH of the wastewater effluent resulting from the dilution and the natural buffering capacity of the wastewater. Changes in wastewater effluent range between 0.0 and 0.2 s.u., on average. As a result, significant changes in the pH of the receiving stream, even under low to no dilution of the wastewater effluent, are expected to be minimal. This has been demonstrated at the laboratory scale (Cavallini & de Souza, 2013; Luukkonen et al., 2014) and at full scale.

5.4 Disinfection Byproducts

Several peer-reviewed studies have investigated the potential formation of disinfection byproducts (DPBs) with the use of PAA in wastewater. Formation of chlorinated or brominated DPBs is not known with use of PAA because PAA is itself not a chlorinated or brominated oxidant. It was shown (Crebelli, Conti, Monarca, Feretti, Zerbini, Zani, Ottavini, 2005) that moderate doses of PAA in the range of 2 to 4 mg/L do not lead to genotoxic DPBs. Dell'Erba, Falsanisi, Liberti, Natarnicola, and Santoro (2004) did show some formation of aldehydes, although at negligible concentrations with respect to the target permit levels for the facility in question. It was shown that PAA was able to further oxidize these aldehydes, and that no halogenated phenols were formed. This is explored further in Chapter 6.

5.5 Addition of Quenching Agents

If it is determined that the PAA residual at the effluent outfall exceeds regulatory compliance, and quenching is appropriate, the addition of quenching reagents may have an impact on downstream water quality, similar to what is expected for hypochlorite quenching. The addition of sodium-based quenching agents, such as sodium bisulfite, will lead to an increase in the sodium content of the effluent and, subsequently, the receiving stream. In addition, if an excess of the reducing agent is added compared to the theoretical amount needed to quench the PAA, the excess reducing agent may lead to deoxygenation of the receiving stream. As a result, excess addition of the reducing agent should be avoided.

6.0 REFERENCES

Baldry, M., Cavadore, A., French, M., Massa, G., Rodrigues, L., Schirch, P., & Threadgold, T. (1995). Effluent disinfection in warm climates with peracetic acid. *Water Science & Technology,* 31, 161–164.

Bringmann, G., & Kuehn, R. (1982). Ergebnisse der Schadwirkung wassergefahrdender Stoffe gegen Daphnia Magna in einem weiterentwickelten standardiisierten Testverfahren. *Wasser Abwasser Forch,* 15, 1–6.

Cavallini, G. S., & de Souza, J. B. (2013). Evaluation of the physical–chemical characteristics of wastewater after disinfection with peracetic acid. *Water, Air, & Soil Pollution,* 224. doi:10.1007/s11270-013-1752-5

Chhetri, R., Baun, A., & Andersen, H. (2017). Algal toxicity of the alternative disinfectants performic acid (PFA), peracetic acid (PAA), chlorine dioxide (ClO_2) and their by-products, hydrogen peroxide (H_2O_2) and chlorite (ClO_2^-). *International Journal of Hygiene and Environmental Health,* 220, 570–574.

Cooper, W., & Lean, D. (1989). Hydrogen peroxide concentration in a norther lake: photochemical formation and diel variability. *Environmental Science & Technology,* 23, 1425–1428.

Crebelli, R., Conti, L., Monarca, S., Feretti, D., Zerbini, I., Zani, C., Ottavini, M. (2005). Genotoxicity of the disinfection by-products resulting from peracetic acid or hypochlorite—disinfected sewage wastewater. *Water Research,* 39, 1105–113.

Degussa, A. G. (1997). Vorversuche zum Fischtest—Ermittlung der kritischen Konzentration an H2O2 in Waser. *ISEGA Industrie—Studien— und Entwicklungs—Goselischaft mbH Aschaffenburg.*

Dell'Erba, A., Falsanisi, D., Liberti, L., Natarnicola, M., & Santoro, D. (2004). Disinfecting behavior of peracetic acid for municipal wastewater reuse. *Desalination,* 168, 435–442.

Doneker, R., & Jirka, G. (1990). *Expert systems for hydrodynamic mixing zone analysis of conventional and toxic single port discharges (CORMIX1)* (EPA/600/3-90/012). Athens, GA: US Environmental Protection Agency, Environmental Response Laboratory Network.

European Centre for Ecotoxicology and Toxicology of Chemicals. (2001). Joint Assessment of Commodity Chemicals (JACC) programme. JACC No. 40, Peracetic acid and its equilibrium solutions. Brussels, Belgium.

European Centre for Ecotoxicology and Toxicology of Chemicals. (1993). Joint Assessment of Commodity Chemicals (JACC) programme. JACC No. 22, Hydrogen peroxide. Brussels, Belgium.

European Chemicals Branch. (2003). European Union risk assessment report, hydrogen peroxide, Vol 38. Retrieved from http://edb.jrc.ec.europa.eu/

documents/existing chemicals/risk_assessments/hydrogenperoixdereport 022.pdf

Frick, W., Roberts, P., Davis, L, Keyes, J, Baumgartner, D., & George, K. (2003). *Dilution models for effluent discharges* (4th ed.) (Visual Plumes). Athens GA: Ecosystems Research Division, NERL, ORD, U.S. Environmental Protection Agency.

Gannon, J., & Gannon, S. (1975). Observations on the narcotization of crustacean zooplankton. *Crustaceana,* 28, 220–225.

Gurtler Industries, Inc. (2015) Safety data sheet, sodium hypochlorite, 12.5%.

Health Canada (2013). *Registration decision: Proxitane.* RD 2013-06.

Henao, L., Compagni, R., Turolla, A., & Antonelli, M. (2018). Influence of inorganic and organic compounds on the decay of peracetic acid in wastewater disinfection. *Chemical Engineering Journal,* 337, 133–142.

Hushangi, R., & Shekarabi, S. (2018). Effect of peracetic acid–based disinfectant on growth, hematology and histology of juvenile rainbow trout (*Oncorchynchus mykiss*). *Fishes,* 3, 1–8.

Kay, S., Quimby, P., & Ouzis, J. (1982). Hydrogen peroxide: A potential algicide for aquaculture. *Proceedings of the Southern Weed Science Society 35th*; Cleveland, MS: Delta State University.

Keogh, B., & Tranh, M. (2011). Old city, new ideas: Peracetic acid in wastewater disinfection in St. Augustine. *Florida Water Resources Journal,* 44–48.

Koivunen, J., & Heinonen-Tanski, H. (2005). Peracetic acid (PAA) disinfection of primary, secondary and tertiary treated municipal wastewaters. *Water Research,* 39, 4445–4453.

Liu, D., Straus, D., Pedersen, L., & Meinelt, T. (2015). Comparison of the toxicity of wofasteril peracetic acid formulations E400, E250, and Lspez to *Daphnia magna,* with emphasis on the effect of hydrogen peroxide. *North American Journal of Aquaculture,* 77, 128–135.

Liu, D., Behrens, S., Pedersen, L., Straus, D., & Menelt, T. (2016). Peracetic acid is a suitable disinfectant for recirculating fish-microalgae integrated multi-trophic aquaculture systems. *Aquaculture Reports,* 4, 136–142.

Liu, D., Straus, D., Pedersen, L., & Meinelt, T. (2018). Periodic bacterial control with peracetic acid in a recirculating aqualcuture system and its long-term effect on fish health. *Aquaculture,* 485, 154–159.

Liu, D. (2018). *Effect of peracetic acid at low concentrations on fish health and water quality.* (published thesis). Liebniz-Institute of Freshwater Ecology and Inland Fisheries, Berlin, Germany. DOI: 10.13140/RG.2.2.23177 .44648

Luukkonen, T., Teeriniemi, J., Proddola, H., & Lassi, U. (2014). Chemical aspects of peracetic acid based wastewater disinfection. *Water SA, 40*, 73–80.

McKee, J. E., & Wolf, H. W. (1963). *Water quality criteria* (2nd ed.). Pasadena, CA: State Water Quality Control Board.

MWH Americas (2014). Peracetic acid effluent discharge mixing evaluation (report to PeroxyChem).

Organization for Economic Co-operation and Development. (2008). Peracetic acid SIDS report. Retrieved from https://hpvchemicals.oecd.org/ui/handler.axd?id=B9C25C3B-98A3-4092-AEEE-9561600F87B9

Organization for Economic Co-operation and Development. (2004). Sodium hypochlorite SIDS report. Retrieved from: http://www.oecd.org/official documents/publicdisplaydocumentpdf/?cote=env/jm/mono(2012)4/part4&doclanguage=en

Pedersen, L., Meinelt, T., & Straus, D. (2013). Peracetic acid degradation in freshwater aquaculture systems and possible practical implications. *Aquacultural Engineering, 53*, 65–71.

PeroxyChem. (2014). *Aquatic toxicity and fate of VigorOx WWT II.*

PeroxyChem. (2015). *7 day renewal chronic toxicity test for VigorOx peracetic acid.* Report to Tennessee Department of the Environment and Conservation.

Pesticide Action Network. (2012). Pesticides database—Chemicals. Retrieved from http://pesticideinfo.org/Detail_Chemical.jsp?Rec_Id=PC34390

Quimby, P. (1961). Preliminary evaluation of hydrogen peroxide as a potential herbicide for aquatic weeds. *Journal of Aquatic Plant Management, 19*, 53–55.

Rossi, S., Antonelli, M., Mezzanotte, V., & Nurizzo, C. (2007). Peracetic acid disinfection: A feasible alternative to wastewater chlorination. *Water Environment Research, 79*, 341–350.

Shurtleff, L. (1989). Sodium percarbonate and hydrogen peroxide—Acute toxicity to the freshwater invertebrate *Daphnia pulex*. Burlington, NC: Interox America.

Southgate, B. (1950). Polluting effects of sewage and industrial waste, toxicity of substances to fish. Report for the Year 1950. *Water Pollution Research Board.*

Stampi, S., de Luca, G., & Zanetti, F. (2001). Evaluation of the efficiency of peracetic acid in the disinfection of sewage effluents. *Journal of Applied Microbiology, 91*, 833–838.

Straus, D., Meinelt, T., Farmer, B., & Beck. B. (2012). Acute toxicity and histopathology of channel catfish fry exposed to peracetic acid. *Aquaculture,* 342, 134–138.

Straus, D., Meinelt, T., Liu, D., & Pedersen, L. (2017). Toxicity of peracetic acid to fish: Variation among species and impact of water chemistry. *Journal of the World Aquaculture Society,* 49, 715–724.

Straus, D. (2018). Introducing a new disinfectant for U.S. aquaculture—Peracetic acid (abstract). Annual Eastern Fish Health Workshop; April 9–13, Chattanooga, Tennessee; p. 22.

Thomann, R. V., & Mueller, J. A. (1987). *Principles of surface water quality modeling and control.* New York, NY: Harper and Row.

U.S. Environmental Protection Agency. (n.d.). E-FAST. Retrieved from https://www.epa.gov/tsca-screening-tools/e-fast-exposure-and-fate-assessment-screening-tool-version-2014

U.S. Environmental Protection Agency. (2000). *Method guidance and recommendations for whole effluent toxicity (WET) testing* (EPA-821/B-00-004). Washington, DC: U.S. Environmental Protection Agency.

U.S. Environmental Protection Agency. (2002). *Methods for measuring the acute toxicity effects of effluents and receiving waters to freshwater and marine organisms* (EPA-821/R-02-012). Washington, DC: U.S. Environmental Protection Agency.

U.S. Environmental Protection Agency. (2010). *Summary of product chemistry, environmental fate, and ecotoxicity data for the chlorine registration review decision document.* Washington, DC: U.S. Environmental Protection Agency.

Water Environment Research Foundation. (2018). *Design and implementation of peracetic acid for wastewater treatment* (WERF Project No. LIFT14T16). Alexandria, VA: Water Environment Research Foundation.

7.0 SUGGESTED READING

Block, P. (2016). The decomposition of peracetic acid and hydrogen peroxide in municipal wastewaters. *Proceedings of the 89th Annual Water Environment Federation Technical Exhibition and Conference;* New Orleans, Louisiana, Sept 24–28; Alexandria, VA: Water Environment Federation.

Risk Sciences. (2001). Regulating whole effluent toxicity using "percent effect" as the test endpoint. Retrieved from http://colowqforum.org/pdfs/whole-effluent-toxicity/03-2010/Percent%20Effect.pdf

6

Peracetic Acid Disinfection Byproducts and Treatment of Contaminants of Concern

Philip Block, Kati Bell, Laura Dominguez Henao, Manuela Antonelli, & Eddie Machek

1.0 INTRODUCTION

Chlorine disinfectants interact with dissolved natural organic matter, inducing the formation of several harmful byproducts. Among these disinfection byproducts (DBPs), trihalomethanes (THMs) and haloacetic acids (HAAs) were the first chlorine DBPs reported (Rook, 1974). The European Union currently regulates THMs, and the United States regulates THMs and HAAs (Richardson, Plewa, Wagner, Schoeny, & DeMarini, 2007). Both are also regulated elsewhere around the world. About 600 DBPs have been identified. Among these are some highly toxic contaminants of concern, such as halonitromethanes, haloacetonitriles, haloamides, halofuranones, iodoacids, and nitrosamines, among others (Deborde & von Gunten, 2008; Richardson et al., 2007).

Given the toxic, mutagenic, and carcinogenic effects that these compounds can have on humans and the environment, DBPs have been recognized as a major problem during the disinfection of both surface water for drinking water and wastewater reuse for irrigation and groundwater replenishing.

In addition to DBPs, other chemical compounds that find their way into drinking and recreational waters may be of concern with respect to impacting human health. Such compounds include human and veterinary pharmaceuticals, plasticizers, industrial chemicals, illicit drugs, personal care products, and organic pesticides (Houtman, 2010). One such class of chemicals are natural and synthetic compounds called endocrine-disrupting compounds (EDCs) that can interrupt or effect the endocrine (hormonal) process by mimicking or blocking the actions of hormones in the body. Exposure to such chemicals may lead to adverse health effects, some of which may take years to decades to become evident. An example of a group of EDCs that have received significant attention are estrogenic compounds that act as the female sex hormone, 17-β-estradiol. These compounds have been implicated in the intersexuality of male fish, which subsequently develop female characteristics (Jobling, Nolan, Tyler, Brighty, & Sumpter, 1998).

Contaminants of concern may enter the drinking water supply through several routes. In addition to incidental spills from manufacturing and other locations, Figure 6.1 shows the most important pathways by which contaminants may be introduced to surface waters and end up in the source water for drinking water production. Runoff from agricultural and farming operations can be a significant source of contaminants of concern, which can include pesticides, phosphates, and nitrates, as well as a source of veterinary pharmaceuticals. Direct discharge from home and industrial sources into

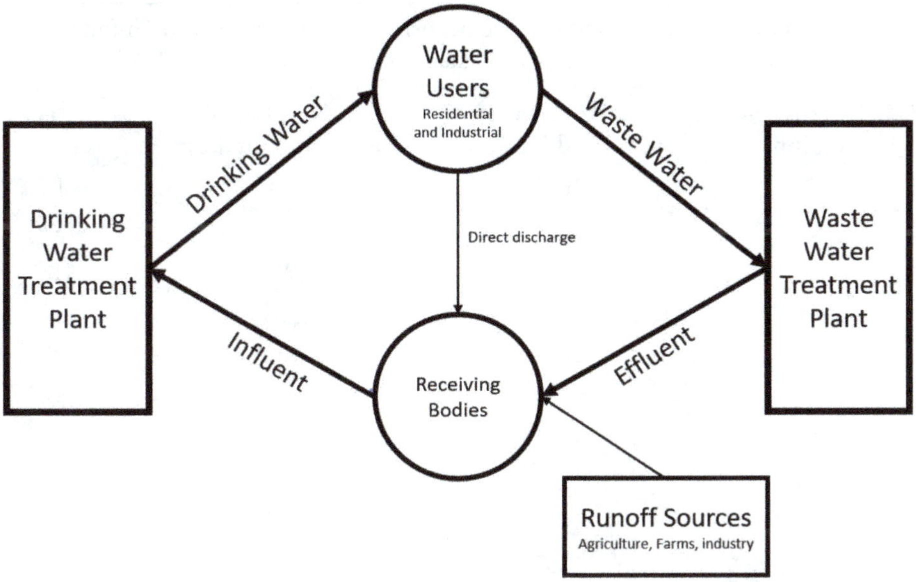

FIGURE 6.1 Water cycle and potential transport routes of CoCs (courtesy of Philip Block).

surface waters also are potential point sources of contaminants entering the environment.

In addition, raw drinking water supplies may be influenced by treated wastewater discharges upstream of the drinking water treatment facility intakes, which is termed "unplanned" or "unintentional indirect potable reuse" or *de facto reuse* (Ngyuen et al., 2018). It is estimated that approximately half of drinking water supplies for populations greater than 10 000 people sourced from surface waters in the United States are affected by at least one upstream treated wastewater discharge (Nguyen et al., 2018; Rice & Westerhoff, 2015). In addition, investigations of drinking water sources downstream of treated wastewater discharges detected pharmaceuticals and other contaminants of concern (Bradley et al., 2017; Glassmeyer et al., 2005).

Rice and Westerhoff (2015) investigated the spatial and temporal variation of de facto water resuse in drinking water systems across the United States. Table 6.1 presents a partial summary of their findings for de facto reuse occurrence under average flow conditions for 14 651 facilities that discharge to surface waters. From this study, it is suggested that contaminants of concern entering a water resource recovery facility (WRRF) that remain intact to the effluent of the facility, or contaminants that may be generated during the treatment process, such as DBPs, may have inadvertent downstream quality impacts on supplied drinking water.

TABLE 6.1 Frequency of de facto reuse occurrence at drinking water intakes across the United States and percent de facto reuse (Rice & Westerhoff, 2015).

U.S. hydrogeologic region	% Number of impacted intakes[a]	% Number of impacted intakes	% de facto reuse
2	36.4	50	<1
5	64.8	89	<5
4	85.1	94	<10
3N	55.1		
18	26.0	21 intakes	>15
6	67.0	16 intakes	>20
11	48.9		
10L	42.9		
17	36.1		
3W	53.7		
7	68.3		
12	89.8		
8	85.0		
3S	44.7		
10U	58.3		
14	52.2		
15	95.5		
13	58.8		
16	6.3		
9	100.0		

[a]Number of drinking water plant intakes that can be impacted by upstream wastewater outflows

2.0 DISINFECTION BYPRODUCTS

2.1 Potential Disinfection Byproduct Formation by Peracetic Acid Disinfection Processes

2.1.1 Carboxylic Acids and Total Organic Carbon

Even though PAA is considered to decompose to oxygen, water, and acetic acid, the possibility that it could form harmful DBPs cannot be completely ignored (Domínguez Henao et al., 2018). Peracetic acid is a strong oxidant, so it can react with organic compounds such as amino acids, phenols, anilines, and other aromatics present in a wastewater effluent, and in turn form aldehydes and carboxylic acids as byproducts (Crathorne et al., 1991; Nurizzo et al., 2005). The carboxylic acid concentrations in the effluent will increase as a result of the addition of PAA because of the acetic acid fraction present in the PAA equilibrium solution and the acetic acid produced during the decomposition of PAA.

Carboxylic acids can also be formed from the oxidation of aqueous organic compounds; however, no formation mechanism has been presented so far for this pathway. A study by Monarca et al. (2002) demonstrated that medium-chain carboxylic acids, such as octanoic, nonanoic, decanoic, lauric, myristic, and hexanedioic acids, were formed after PAA (1.5 and 3.0 mg/L) disinfection of surface water. However, these carboxylic acids are not considered toxic, mutagenic, or carcinogenic.

Consequently, PAA increases the total organic carbon (TOC) level as well in treated effluents. In several studies, an addition of 5 mg/L PAA to wastewater increases the TOC level by a factor of 1.5, and an addition of 10 mg/L triples the TOC level (Profaizer, 1998; U.S. EPA, 2007; Wagner, Brumelis, & Gehr, 2002).

2.1.2 Aldehyde Formation During Peracetic Acid Disinfection

Modest concentrations of aldehydes (10 to 30 µg/L) have been found after PAA disinfection of municipal wastewater arising from radical pathways when PAA interacted with amino acids, phenols, and other aromatic substances present in treated wastewater (Crathorne et al., 1991; Kitis, 2004). Experimental results (Dell'Erba, Falsanisi, Liberti, Notarnicola, & Santoro, 2007; Liberti & Notarnicola, 1999; Monarca et al., 2002; Nurizzo et al., 2005) have evidenced that a limited formation of aldehydes can occur, directly proportional to PAA dosage. Aldehyde formation varies with PAA/COD ratio, reaching a maximum around 0.3 mg/mg and then decreases at higher ratios (Dell'Erba et al., 2007). This is a resulting consequence, that once formed, aldehydes can be further oxidized by excess PAA to the corresponding carboxylic acids, or ultimately to carbon dioxide (CO_2) according to the following reactions, as observed by Dell'Erba et al. (2007):

$$2\ CH_3COOOH + CH_3OH \rightarrow 2\ CH_3COOH + HCHO + H_2O \quad (6.1)$$

$$CH_3COOOH + HCHO \rightarrow 2\ CH_3COOH + CO_2 + H_2O \quad (6.2)$$

The aldehydes formed after PAA treatment included formaldehyde, acetaldehyde, nonanal, and decanal, among others (Monarca et al., 2002). However, Cavallini et al. (2013) did not observe any byproducts such as aldehydes, phenols, and benzene derivatives during the disinfection of secondary effluents with PAA (at a level of 10 mg/L).

Regarding regulatory aspects, the maximum allowable concentrations of aldehydes in Italy are 1 mg/L (as formaldehyde, HCHO) for wastewater discharge in surface water (Ministry of the Environment of Italy, 2006a) and 0.5 mg/L (as HCHO) for wastewater reuse (Ministry of the Environment of Italy, 2006b), respectively. The formation of aldehydes observed by Italian researchers has been limited and never exceeded these limits.

As for U.S regulations, the U.S. Environmental Protection Agency (U.S. EPA) has evaluated formaldehyde and acetaldehyde for carcinogenicity. The chemical assessment summaries published on the Integrated Risk Information System (IRIS) for formaldehyde (U.S. EPA, 1989) and acetaldehyde (U.S. EPA, 1988) classified them as probable human carcinogens in the categories B1 and B2, respectively, based on animal data and some limited human data from occupational exposure. However, these evaluations did not include quantitative risk estimates for cancer published on the IRIS for aldehydes such as formaldehyde and acetaldehyde. Neither of these two assessments on IRIS has been revised since 1990 (Richardson et al., 2007).

2.1.3 Halogenated Disinfection Byproducts

The formation of halogenated DBPs can occur at high PAA concentrations through the oxidation of halide ions to form hypohalous acids, which subsequently react with dissolved organic matter to form halogenated DBPs (Dell'Erba et al., 2007; Luukkonen & Pehkonen, 2017; Shah et al., 2015b). Monarca, Zani, Richardson, Thruston, Moretti, Feretti, and Villarini (2004) observed a very limited formation of halogenated organic DBPs after PAA treatment of surface water as the sum parameter of adsorbable organic halogens; however, no specific compounds were identified. Pedersen, Meinelt, and Straus (2013) also studied the formation of byproducts after PAA disinfection at concentrations between 0 to 4.8 mg/L and contact times from 6 to 37 minutes. The analyses of the bromophenols, chlorophenols, and formaldehyde were below 0.1 µg/L, the detection limit of the analysis methods, which is in agreement with Dell'Erba et al. (2007).

The electrochemistry of PAA enables the oxidation of bromide to hypobromous acid, but not of chloride to hypochlorous acid. Indeed, high concentrations of PAA (30 mg/L) can oxidize bromide to hypobromous acid (HOBr) in solutions enriched with bromides and phenol (400 µg/L) and extensive contact time (14 hours), forming 2- and 4-bromophenol (Booth & Lester, 1995; Crathorne et al., 1991; Wagner et al., 2002). This suggests that PAA can lead to the formation of brominated byproducts, albeit under unrealistic conditions in conventional drinking or wastewater treatment.

High concentrations of PAA or halide anions boost the formation of halogenated DBPs, as observed by Booth and Lester (1995). This was also found during the disinfection of saline ballast water (Shah et al., 2015a). However, under typical disinfection conditions of surface or wastewater with PAA, where typical PAA dose concentrations are below 5 mg/L, only very limited or undetectable amounts of chlorinated and brominated DBPs (e.g., bromate, chlorophenols, bromophenols, or total trihalomethanes) have been observed (Dell'Erba et al., 2007; Liberti & Notarnicola, 1999; Monarca et al., 2002, 2004; Nurizzo et al., 2005). For instance, the disinfection of sea or brackish water with high concentrations of PAA (150 mg/L) and extended contact time (of 5 days) was able to form bromoform (up to 920 µg/L) and dibromochloromethane (up to 110 µg/L), whereas no formation of chloroform or bromodichloromethane was observed (Luukkonen & Pehkonen, 2017; Shah et al., 2015a). No chlorinated DBPs were reportedly observed for PAA-treated water (Liberti & Notarnicola, 1999; Luukkonen & Pehkonen, 2017; Richardson et al., 2007).

2.1.4 Potential for Epoxide Formation

Although PAA is a common epoxidation agent for alkenes in the chemical industry (Kaur & Kishore, 2014), no formation of epoxides was observed as a result of PAA disinfection of wastewater (10 mg/L, 30 minutes) (Liberti & Notarnicola, 1999). Indeed, epoxides are very unstable and, if formed, should behave as reaction-intermediates that immediately decompose to hydrogen peroxide (H_2O_2) and carbonyl-containing products.

2.1.5 Nitrogenous Disinfection Byproducts

The awareness of contaminants of concern (CoCs) has shifted the research of DBPs toward nitrogenous disinfection byproducts (N-DBPs) (Shah & Mitch, 2012), which include halonitroalkanes, halonitriles, haloamides, and N-nitrosamines. In particular, the N-nitrosamines represent a group of emerging DBPs that has received increasing attention for being more carcinogenic and mutagenic than THMs and HAAs, even if they are formed at much lower concentrations. The most notable one, N-nitrosodimethylamine (NDMA), has an associated cancer risk approximately 600 times greater than any of the regulated THMs, with a risk at concentrations as low as 0.7 ng/L in drinking water (Lee, Westerhoff, & Croué, 2007; Shah &

Mitch, 2012; West et al., 2016). West, Wu, Donovan, Shi, Ma, Jiang, and Wang (2016) investigated the formation of eight N-nitrosamines in both the presence and absence of amine-based N-nitrosamine precursors during disinfection with PAA, free chlorine, and monochloramine in drinking water. In the presence of precursors, N-nitrosamine formation was observed by monochloramine and free chlorine disinfection, whereas PAA disinfection did not produce NDMA or any other N-nitrosamines above detection limits, except low concentrations of N-nitrosodi-n-propylamine (NDPA) ($<$10 ng/L) when precursors were added in the water samples.

2.1.6 Endocrine Disruptors

Peracetic acid is not included in the European Union's *Commission Staff Working Document on implementation of the Community Strategy for Endocrine Disrupters* that establishes a range of substances suspected of interfering with the hormone systems of humans and wildlife (Commission to the European Parliament and the Council, 2016). There is no evidence of any endocrine-disruption potential in the human health or ecotoxicological studies. As discussed in Section 3.2, research indicates that PAA may reduce the endocrine disruption potential of certain wastewaters.

2.2 The Influence of Hydrogen Peroxide on Disinfection Byproduct Formation

The effect of the hydrogen peroxide (H_2O_2) fraction in equilibrium with PAA and its effect on the formation of byproducts during PAA disinfection has not been investigated. Nevertheless, it has been observed that it has an influence in the formation DBPs during disinfection of saline waters. Indeed, H_2O_2 can exhibit oxidizing and reducing properties, depending on the pH, and in basic solutions H_2O_2 can reduce inorganic ions. When saline waters were treated with PAA solutions with high molar ratio H_2O_2:PAA (150 mg PAA/L, contact time 18 to 24 hours), H_2O_2 was able to reduce HOCl, HOBr, or HOI to chloride, bromide, or iodide respectively, leading to a limited formation of halogenated DBPs. In contrast, when disinfection is performed with PAA formulations with low molar ratio H_2O_2:PAA, significant concentrations of brominated DBPs, such as bromoform (260 µg/L), bromoacetic acid (106 µg/L), dibromoacetic acid (230 µg/L), and tribromoacetic acid (89 µg/L), were formed (Shah et al., 2015b).

Thus, H_2O_2 can potentially serve as an important sink for secondary oxidants such as hypohalous acids and minimize their further reaction with organic matter to form THMs and HAAs. Up to now, the implications of H_2O_2 during PAA treatment have not been investigated. In other oxidation systems (e.g., O_3/H_2O_2), H_2O_2 reduced the HOBr concentration and, thereby, bromate formation (Shah et al., 2015b).

3.0 PERACETIC ACID DEGRADATON OF CONTAMINANTS OF CONCERN

3.1 Direct Oxidation

Peracetic acid has been implemented as an oxidant for over 30 years. However, the research on the ability of PAA to degrade contaminants of concern (CoCs) is fairly limited. Some examples of experiments that were conducted for the direct oxidation of CoCs by PAA include antibiotics, anti-inflamatories, lipid regulating agents, and other CoCs (Hey, Ledin, la cour Jansen, & Andersen, 2012; Zhang et al., 2017). While certain moieties were able to be oxidized by PAA directly, many compounds proved to be inert unless advanced oxidation processes (AOPs) involving PAA were applied. The selectivity of PAA is surprising because of its high oxidation potential of 1.8 V.

Table 6.2 displays reaction rates and contaminant reduction percentages for a number of contaminants of concern, including pharmaceutical compounds.

3.2 Peracetic Acid Oxidation of Endocrine-Disrupting Compounds

In addition to providing effective disinfection, recent research has focused on evaluating this technology for improving the characteristics of effluent with respect to oxidizing EDCs. Recent unpublished work (Block, Reimers, & Xu, 2015) demonstrated the effectiveness of PAA in reducing endocrine disrupting activity in effluent from a WRRF in Louisiana. The effluent from this facility is being considered for discharge in the surrounding wetlands to help restore the wetlands following the impact of Hurricane Katrina. The facility currently uses hypochlorite as its disinfection technology and, as a result, cannot currently discharge its wastewater into the wetlands because of the formation of DBPs.

As a result, the facility investigated PAA as an alternative disinfectant to aid in protecting sensitive species and to allow for discharge into the wetlands. Testing was performed at laboratory scale and using a pilot reactor located near the existing chlorine contact chamber. While the primary goal of PAA addition was to meet bacterial discharge requirements of <200 CFU/100 mL of fecal coliforms, additional measurements of

TABLE 6.2 Reported oxidation rates/concentrations of contaminants of concern with PAA.

Contaminants of concern	Oxidation system	Water type	First order rate constant (1/min)	Second order rate constant (1/Ms)	Reference
amoxicillin	PAA	Buffered laboratory water	—	28.24	Zhang et al., 2017
ampicillin	PAA	Buffered laboratory water	—	28.77	Zhang et al., 2017
bezafibrate	PAA/UV	Deionized water	2.4×10^{-2}	—	Cai et al., 2017
carbamazepine	PAA/UV	Deionized water	8.25×10^{-2}	—	Cai et al., 2017
cefadroxil	PAA	Buffered laboratory water	—	44.38	Zhang et al., 2017
cefalexin	PAA	Buffered laboratory water	—	41.43	Zhang et al., 2017
cefapirin	PAA	Buffered laboratory water	—	15.56	Zhang et al., 2017
cephalothin	PAA	Buffered laboratory water	—	17.44	Zhang et al., 2017
clofibric acid	PAA/UV	Deionized water	2.3×10^{-1}	—	Cai et al., 2017
diclofenac	PAA/UV	Deionized water	1.38	—	Cai et al., 2017
ibuprofen	PAA/UV	Deionized water	8.83×10^{-2}	—	Cai et al., 2017
naproxen	PAA/UV	Deionized water	3.07×10^{-1}	—	Cai et al., 2017
penicillin	PAA	Buffered laboratory water	—	19.04	Zhang et al., 2017
		Max removal (%)	PAA dose (mg/L)	Contact time (min)	
4-chlorophenol	PAA/UV	98	3040	1	Sharma, 2010
4-chlorophenol	PAA/UV	95	2040	30	Daswat & Mukhopadhyay, 2014
dichlorophenol	PAA/ Fe_2O_3	70	20 000	210	Virkutyte & Varma, 2014
chlorophenols	PAA/UV	97	4035	30	Daswat & Mukhopadhyay, 2012
phenol	PAA/ MnO_2	80	50	120	Rokhina, Makarova, Golovina, van As, & Virkutyte, 2010
bisphenol-A	PAA/ Fe^{+2}	100	40	10	Luukkonen et al., 2015
Pharmaceuticals	PAA	0–90	15–50	1,080	Hey et al., 2012

estrogenecity were also collected on non-disinfected effluent as well as PAA and chlorine-treated effluent. Endocrine activity of samples was measured by extracting samples at Tulane University and measuring the extract with a receptor gene method (Pons, Gagne, Nicolas, & Mehtali, 1990) at Xavier University. The results showed that PAA reduces the endocrine disrupting activity of the wastewater; compared to chlorine, which was shown to increase endocrine disrupting activity, there is a significant improvement in effluent quality that could have a significant positive impact on receiving water ecology.

3.3 Peracetic Acid as a Component of an Advanced Oxidation System

Enhancement of the degradation capabilities of PAA can be achieved through advanced oxidation processes (AOPs). A key advantage of a PAA AOP system over an H_2O_2 AOP system (such as UV combined with hydrogen peroxide) is that the PAA peroxy bond (38 kcal/mol) is weaker than the H_2O_2 peroxy bond (51 kcal/mol) (Luukkonen & Pehknonen, 2017). Also, many of the compounds studied so far have shown that direct oxidation contributes a negligible amount for removal of CoCs when compared to activated PAA.

3.3.1 Peracetic Acid with Ultraviolet Advanced Oxidation Process

Figure 6.2 shows the potential radical formation during a PAA + ultraviolet (UV) AOP. It has been demonstrated that PAA/UV is comparable in price to H_2O_2/UV in achieving disinfection (Lubello, Caretti, & Gori, 2002). This makes the PAA/UV process interesting for CoC degradation because not only does the PAA solution contain the reaction pathways that H_2O_2/UV would have, but it also has other radical pathways that include the alcoxyl

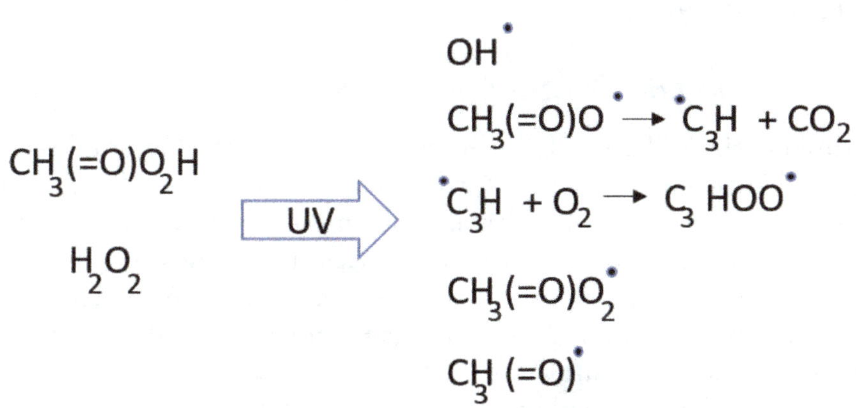

FIGURE 6.2 Peracetic acid/UV AOP radical formation.

and methyl radicals. This means additional compounds that are recalcitrant toward hydroxyl radicals may be degraded by the additional radicals unique to the PAA/UV system (Flores et al., 2014; Rohkina et al., 2010), as shown in Figure 6.2.

Degradation of 4-chlorophenol was achieved in a UV/PAA AOP experiment using 3040 mg PAA/L, yielding 95% removal of the contaminant (Mukhopadhyay & Daswat, 2012). Similary, as seen in Table 6.2, phenols and a variety of pharmaceticals can be reduced by UV/PAA under a variety of PAA dose concentrations and contact times.

3.3.2 Transition Metal Catalyzed Activation of Peracetic Acid

Transition metals and transition metal oxides have been used as heterogeneous catalysts for hydrogen peroxide and PAA reactions (Luukkonen & Pekhonen, 2017). Cuprous and ferrous ions, Cu^{2+} and Fe^{2+}, have been tested to determine their impact on bisphenol-A (BPA). Without a catalyst, no degradation occurred within 60 minutes with PAA dosed at 20 mg/L (Luukkonen et al., 2015). However, once the pH was lowered to 3.5 and the transition metals added, BPA degraded by more than 90% in a few minutes when exposed to commercial PAA (Luukkonen et al., 2015). Another set of experiments was conducted to observe the degradation of Orange II dye in the presence of manganese and PAA. These experiments were carried out at pH 9.5 and demonstrated that PAA alone was not sufficient to degrade the dye. However, the dye was almost completely degraded in 2 minutes with 5-mM PAA and 10^{-5} M Mn(II) catalyst (Rothbart, Ember, & van Eldik, 2012).

Several experiments have been conducted with transition metal oxides as catalysts in PAA systems. One study has found that phenol could be degraded by 80% with a 50 ppm PAA solution and 0.7 g/L of manganese dioxide (MnO_2) catalyst (Rokhina et al., 2010). Interestingly, both catalyst and PAA showed no reactivity towards phenol independent of each other (Rokhina et al., 2010).

3.3.3 Peracetic Acid with Activated Carbon Fibers

Activated carbon fibers (ACFs), in combination with PAA, are an environmentally friendly alternative for the activation of PAA (Zhou, Lu, Yao, Sun, Gong, Li, Pei, . . . Chen, 2015). The compounds investigated were the organic dyes Reactive Brilliant Red X-3B, Acid Red 1, Acid Orange 7, and Basic Green 1 (Zhou et al., 2015). The theory behind the ACF activation of PAA, and subsequent degradation of organic molecules, is that the oxygen-containing sites at the surface of ACFs act as electron transfer mediators that lead to the radical pathways for the degradation of the organic dyes (Zhou et al., 2015).

4.0 SUMMARY AND CONSIDERATIONS

4.1 Implications for Contaminants of Concern Reduction in Water Resource Recovery Facility Schemes

The interest in PAA as an alternatve disinfectant for municpal and industrial wastewaters lies in its ability to kill microbes while not generating chlorinated byproducts. Peracetic acid rapidly breaks down into oxygen, water, and acetic acid, which generate low aquatic toxicity. And while the formation of nonregulated disinfection byproducts from the use of PAA cannot be dismissed, its potential to form harmful, toxicological residuals is low compared to chlorination and other alternative disinfectants. As a result, one of the significant drivers of the market acceptance of PAA in the wastewater industry is its lack of THM and HAA formation potential.

As federal and state regulatory guidelines tighten on the allowable maximum concentrations of THMs, HAAs, bromates, and chlorites, and as potentially problematic compounds such as iodo-organic complexes are under review for regulation, PAA can be a potential tool to address permit requirements for newly designed or renovated WRRFs. In addition, PAA may be a useful disinfectant techology for treatment facilities whose outfall is into senstive aquatic environments that cannot tolerate EDCs (wetlands being one example).

4.2 Implications for Potable Reuse

Contaminants of concern that either pass through or are generated within a WRRF may impact downstream drinking water treatment facilities through either de facto or planned reuse of that water. Unless untreated, these contaminants may pass into the drinking water service system and represent potential sources of risk to human health. One example is unused or unprocessed pharmaceuticals or biologics that pass into the wastewater treatment process and survive to reach the facilities' effluent and are cycled into local surface waters. These waters can then be taken up at the drinking water facility inlets and eventually make their way to the drinking water tap. Although, the World Health Organization has found that, in general, concentrations of pharmaceuticals in drinking water are typically a thousand-fold below the minimum therapuetic dose (World Health Organization, 2011).

Not only can PAA reduce the potential for formation of chlorinated disinfection byproducts during the disinfection of wastewaters, thereby reducing the potential overall concentraiton of DBPs, PAA represents a potential technology that can be used to help ameliorate DPBs and CoCs that may enter a drinking water treatment facility through reuse avenues, whether planned or incidental. Either through direct oxidiation or as part of an advanced oxidation process, PAA can be deployed pre-disinfection in a drinking water facility.

5.0 REFERENCES

Block, P., Reimers, R., & Xu, Y. (2015). Use of peracetic acid as a wastewater disinfectant to eliminate the formation of chlorinated disinfection by-products and inhibit the activity of endocrine disrupting compounds. *Water Environmental Federation Technology Symposium*; New Orleans, LA.

Booth, R. A., & Lester, J. N. (1995). The potential formation of halogenated by-products during peracetic acid treatment of final sewage effluent. *Water Research,* 29, 1793–1801.

Bradley, P., Journey, C., Romanonk, K. Barber, L, Buxton, H., Forman, W., & Furlong, E. (2017). Comprehensive target-chemical assessment reveals extensive mixed-organic contaminant exposure in USA streams. *Environmental Science & Technology,* 9, 4792–4802.

Cai, M., Sun, P., Zhang, L., & Huang, C. (2017). UV/Peracetic acid for degradation of pharmaceuticals and reactive species evaluation. *Environmental Science & Technology,* 51, 14217–14224.

Cavallini, G. S., de Campos, S. X., de Souza, J. B., & de Sousa Vidal, C. M. (2013). Evaluation of the physical-chemical characteristics of wastewater after disinfection with peracetic acid. *Water, Air, & Soil Pollution,* 224.

Commission to the European Parliament and the Council. (2016). Commission Staff Working Document. Impact Assessment. Defining criteria for identifying endocrine disruptors in the context of the implementation of the plant protection products regulation and biocidal products regulation. Brussels, Belgium.

Crathorne, B., Fawell, J., Irving, T., Harris, N., Denny, S., Whitmore, T., Horth, H., . . . Taylor, L. (1991). *Sewage disinfection: by-product formation, ecotoxicology and microbiological efficacy* (Report NR 2727). Medmenham, UK: Water Research Centre.

Daswat, D. P., & Mukhopadhyay, M. (2012). Photochemical degradation of chlorophenol industry wastewater using peroxy acetic acid (PAA). *Chemical Engineering Journal,* 209, 1–6.

Daswat, D. P., & Mukhopadhyay, M, (2014). Effect of UV input on degradation of 4–chlorophenol by peroxy acetic acid. *Arabian Journal for Science and Engineering,* 39, 5873–5881.

Deborde, M., & von Gunten, U. (2008). Reactions of chlorine with inorganic and organic compounds during water treatment—Kinetics and mechanisms: A critical review. *Water Research,* 42, 13–51.

Dell'Erba, A., Falsanisi, D., Liberti, L., Notarnicola, M., & Santoro, D. (2007). Disinfection by-products formation during wastewater disinfection with peracetic acid. *Desalination,* 215, 177–186.

Domínguez Henao, L., Turolla, A., & Antonelli, M. (2018). Disinfection by-products formation and ecotoxicological effects of effluents treated with peracetic acid: A review. *Chemosphere*, 213, 25–40.

Flores, M., Lescano, M., Brandi, R., Cassano, A., & Labas, M. (2014). A novel approach to explain the inactivation mechanism of Escherichia coli employing a commercially available peracetic acid. *Water Science & Technology,* *69*(2), 358–363.

Glassmeyer, S., Furlong, E., Kolpin, D., Cahill, J., Zaugg, S., Werner, S., Meyer, M., & Kryak, D. (2005). Transport of chemical and microbial compounds from known wastewater discharges: Potential for use as indicators of human fecal coontamination. *Environmental Science & Technology,* *39*(14).

Hey, G., Ledin, A., Jansen, L., Andersen, H. (2012) Removal of pharmaceuticals in biologically treated wastewater by chlorine dioxide or peracetic acid. *Environmental Technology,* *33*(9), 1041–1047.

Houtman, C. J. (2010). Emerging contaminants in surface waters and their relavance for the production of drinking water in Europe. *Journal of Integrative Environmental Sciences,* 7(4), 271–295.

Jobling, S., Nolan, M., Tyler, C., Brighty, G., & Sumpter, J. (1998). Widespread sexual disruption in wold fish. *Environmental Science & Technology,* 32, 2398–2506.

Kaur, N., & Kishore, D. (2014). Peroxy acids: Role in organic synthesis. *Synthetic Communications,* 44, 721–747.

Kitis, M. (2004). Disinfection of wastewater with peracetic acid: A review. *Environment International,* 30, 47–55.

Lee, W., Westerhoff, P., & Croué, J.-P. (2007). Dissolved organic nitrogen as a precursor for chloroform, dichloroacetonitrile, N-nitrosodimethylamine, and trichloronitromethane. *Environmental Science & Technology,* 41, 5485–90.

Liberti, L., & Notarnicola, M. (1999). Advanced treatment and disinfection for municipal wastewater reuse in agriculture. *Water Science & Technology,* 40, 235–245.

Lubello, C., Caretti, C., & Gori, R. (2002). Comparison between PAA/UV and H2O2/UV disinfection for wastewater reuse. *Water Science & Technology: Water Supply, 2*(1), 205–212.

Luukkonen, T., Heyninck, T., Ramo, J., & Lassi, U. (2015). Comparison of organic peracids in wastewater treatment: Disinfection, oxidation and corrosion. *Water Research,* 85, 275–285.

Luukkonen, T., & Pehkonen, S. O. (2017). Peracids in water treatment: A critical review. *Critical Reviews in Environmental Science & Technology, 47*(1), 1–39.

Ministry of the Environment of Italy (2006a). Decree 152/06, Gazzetta Ufficiale della Reppublica Italiana, No. 88, 14 April 2006 (in Italian). Italy.

Ministry of the Environment of Italy (2006b). Decree n. 168, Gazzetta Ufficiale della Reppublica Italiana, No.108, 11 May 2006 (in Italian). Italy.

Monarca, S., Richardson, S. D., Feretti, D., Grottolo, M., Thruston, A. D., Zani, C., Navazio, G., . . . Alberti, A. (2002). Mutagenicity and disinfection by-products in surface drinking water disinfected with peracetic acid. *Environmental Toxicology & Chemistry,* 21, 309–318.

Monarca, S., Zani, C., Richardson, S. D., Thruston, A. D., Moretti, M., Feretti, D., & Villarini, M. (2004). A new approach to evaluating the toxicity and genotoxicity of disinfected drinking water. *Water Research,* 38, 3809–3819.

Mukhopadhyay, M., & Daswat, D. P. (2012). Photochemical degradation of 4-chlorophenol in the aqueous phase using peroxyacetic acid (PAA). *Water Science & Technology, 67*(2), 440–45.

Nguyen, T., Westerhoff, P., Furlong, E., Koplin, D., Batt, A., Mash, H., Schenck, K., . . . Glassmeyer, S. (2018). Modeled de facto reuse and contaminants of concern in drinking water source waters. *Journal of the American Water Works Association, 110*(4), E2–E18.

Nurizzo, C., Antonelli, M., Profaizer, M., & Romele, L. (2005). By-products in surface and reclaimed water disinfected with various agents. *Desalination,* 176, 241–253.

Pedersen, L. F., Meinelt, T., & Straus, D. L. (2013). Peracetic acid degradation in freshwater aquaculture systems and possible practical implications. *Aquacultural Engineering,* 53, 65–71.

Pons, M., Gagne, D., Nicolas, J. C., & Mehtali, M. (1990). A new cellular model of response to estrogens: A bioluminescent test to characterize (anti) estrogen molecules. *Biotechniques,* 9, 450–459.

Profaizer, M. (1998). *Peracetic acid as an alternative technology for drinking water disinfection: A new modelling approach* (Doctoral dissertation) (in Italian). Politecnico di Milano, Milan, Italy.

Rice, J., & Westerhoff, P. (2015). Spatial and temporal variation in de facto wastewater reuse in drinking water systems across the U.S.A. *Environmental Science & Technology,* 49, 982–989.

Richardson, S. D., Plewa, M. J., Wagner, E. D., Schoeny, R., & DeMarini, D. M. (2007). Occurrence, genotoxicity, and carcinogenicity of regulated and emerging disinfection by-products in drinking water: A review and roadmap for research. *Mutation Research,* 636, 178–242.

Rokhina, E. V., Makarova, K., Golovina, E. A., van As, H., & Virkutyte, J. (2010). Free radical reaction pathway, thermochemistry of peracetic acid homolysis, and its application for phenol degradation: Spectroscopic study and quantum chemistry calculations. *Environmental Science & Technology,* 44, 6815–6821.

Rook, J. J. (1974). Formation of haloforms during chlorination of natural waters. *Water Treatment and Examination,* 23, 234–243.

Rothbart, S., Ember, E. E., & van Eldik, R. (2012). Mechanistic studies on the oxidative degradation of Orange II by peracetic acid catalyzed by simple manganese (II) salts. Tuning the lifetime of the catalyst. *New Journal of Chemistry,* 36.3, 732–748.

Shah, A. D., Liu, Z.-Q., Salhi, E., Höfer, T., Werschkun, B., & von Gunten, U. (2015a). Formation of disinfection by-products during ballast water treatment with ozone, chlorine, and peracetic acid: Influence of water quality parameters. *Environmental Science: Water Research & Technology,* 1, 465–465.

Shah, A. D., Liu, Z. Q., Salhi, E., Höfer, T., & von Gunten, U. (2015b). Peracetic acid oxidation of saline waters in the absence and presence of H2O2: Secondary oxidant and disinfection byproduct formation. *Environmental Science & Technology,* 49, 1698–1705.

Shah, A. D., & Mitch, W. A. (2012). Halonitroalkanes, halonitriles, haloamides, and N-nitrosamines: A critical review of nitrogenous disinfection byproduct formation pathways. *Environmental Science & Technology,* 46, 119–131.

Sharma, S. (2010). Degradation of 4–chlorophenol in wastewater by organic oxidants. *Industrial & Engineering Chemistry Research,* 49, 3094–3098.

Soares Cavallini, G., Campos, S. X. De, Beber De Souza, J., & Sousa Vidal, C. M. De. (2013). Evaluation of the Physical-Chemical Characteristics

U.S. Environmental Protection Agency. (1988). *Acetaldehyde. Chemical Assessment Summary. Integrated Risk Information System (IRIS).* Washington, DC: U.S. Environmental Protection Association, National Center for Environmental Assessment.

U.S. Environmental Protection Agency. (1989). *Formaldehyde. Chemical Assessment Summary. Integrated Risk Information System (IRIS)*. Washington, DC: U.S. Environmental Protection Agency, National Center for Environmental Assessment.

U.S. Environmental Protection Agency. (2007). Summary Review of Available Literature for Hydrogen Peroxide and Peroxyacetic Acid for New Use to Treat Wastewater. Washington, DC: U.S. Environmental Protection Agency, Office of Prevention, Pesticides and Toxic Substances.

Virkutyte, J., & Varma, R. S. (2014) Eco-friendly magnetic iron oxide–pillared montmorillonite for advanced catalytic degradation of dichlorophenol. *ACS Sustainable Chemistry & Engineering,* 2, 1545–1550.

Wagner, M., Brumelis, D., & Gehr, R. (2002). Disinfection of wastewater by hydrogen peroxide or peracetic acid: Development of procedures for measurement of residual disinfectant and application to a physicochemically treated municipal effluent. *Water Environment Research,* 74, 33–50.

West, D. M., Wu, Q., Donovan, A., Shi, H., Ma, Y., Jiang, H., & Wang, J. (2016). N-nitrosamine formation by monochloramine, free chlorine, and peracetic acid disinfection with presence of amine precursors in drinking water system. *Chemosphere,* 153, 521–527.

World Health Organization. (2011). Pharmaceuticals in drinking water (WHO/HSE/WSH/11.05).

Zhang, K., Zhou, X., Du., P., Zhang, T., Cai, M., Sun, P., & Huang, C. (2017). Oxidation of β-lactam antibiotics by peracetic acid: Reaction kinetics, product and pathway evaluation. *Water Research,* 123, 153–161.

Zhou, F., Lu, C.; Yao, Y., Sun, L., Gong, F., Li, D., Pei, K., Lu, W., & Chen, W. (2015). Activated Carbon Fibers as an Effective Metal-free Catalyst for Peracetic Acid Activation: Implications for the Removal of Organic Pollutants. *Chemical Engineering Journal,* 281, 953–960.

6.0 SUGGESTED READING

Guilherme, S., & Rodriguez, M. Occurrence of regulated and non-regulated disinfection by-products in small drinking water systems, *Chemosphere, 117*(204), 435–432.

Kirmeyer, G. J., Foust, G. W., Pierson, G. L., Simmler, J. J., & LeChevallier, M. W. (1993). *Optimizing chloramine treatment* (Report No. 90620). Denver, CO: American Water Works Association Research Foundation.

Mendez-Arriaga, F., Esplugas, S., & Gimenez, J. (2010). Degradation of the emerging contaminant ibuprofen in water by photo-Fenton. *Water Research, 44*(2), 589–595.

Nikolaou, A., Golfinopoulos, S., Lekkas, T., & Arhonditsis, G. (2004). Factors affecting the formation of organic by-products during water chlorination: A bench scale study. *Water, Air, & Soil Pollution,* 159, 357–371.

Pereira, V. J., Weinberg, H., Linden, K., & Singer, P. (2007). UV degradation kinetics and modeling of pharmaceutical compounds in laboratory grade and surface water via direct and indirect photolysis at 254 nm. *Environmental Science & Technology, 41*(5), 1682–1688.

Plewa, M., & Wagner, E. (2009). *Mammalian cell cytotoxicity and genotoxicity of disinfection by-products* (Project #3089). Denver, CO: Water Research Foundation.

Richardson, S. D. (2003). Disinfection by-products and other emerging contaminants in drinking water. *TrAC Trends in Analytical Chemistry,* 22, 666–684.

Richardson, S. D., Simmons, J. E., & Rice, G. (2002). Disinfection byproducts: The next generation. *Environmental Science & Technology,* 36, 198A–205A.

Rokhina, E. V., Makarova, K., Lahtinen, M., Golovina, E., Van As, H., &Vinkutyte, J. (2013). Ultrasound-assisted MnO 2 catalyzed homolysis of peracetic acid for phenol degradation: The assessment of process chemistry and kinetics. *Chemical Engineering Journal,* 221, 476–486.

Schnitzer, M., & Hindle, D. A. (1980). Effect of peracetic acid oxidation on N-containing components of humic materials. *Canadian Journal of Soil Science, 60*(3), 541–548.

U.S. Environmental Protection Agency. (1998). *National Primary Drinking Water Regulations: Disinfectants and Disinfection Byproducts* (Vol. 63, #241, 40 CFR parts 9, 141 and 142). Washington, DC: U.S. Environmental Protection Agency.

U.S. Environmental Protection Agency. (2005). *Technology and costs document for the Final Long Term 2 Enhanced Surface Water Treatment Rule and Final State 2 Disinfectants and Disinfectants Byproducts Rule* document (EPA 815-R/05-013). Washington, DC: U.S. Environmental Protection Agency.

Wols, B. A., & Hofman-Caris, C. H. M. (2012). Review of photochemical reaction constants of organic micropollutants required for UV advanced oxidation processes in water. *Water Research, 46*(9), 2815–2827.

Xagoraraki, I., & Kuo, D. (2008). Water pollution: Emerging contaminants associated with drinking water. *International Encyclopedia of Public Health*, 1st ed., vol. 6, 539–550.

7

Peracetic Acid Performance Testing

Denise Funk, Edyta Stec-Uddin, Daniela Castañeda, Amy Borello Gruss, & Kati Bell

1.0 INTRODUCTION TO A PHASED EVALUATION APPROACH

To successfully develop information to evaluate peracetic acid (PAA) as a wastewater disinfectant, performance testing should be conducted to assess site-specific applications. This chapter covers PAA application to biologically treated effluent (secondary/tertiary treatment); it does not cover PAA application to primary effluent or combined sewer overflow (CSO) with substantially different water quality characteristics. Performance testing helps determine the PAA dose needed to meet permit compliance requirements for microbial treatment objectives. Site-specific evaluation is particularly helpful in optimizing PAA dosing, both in terms of disinfection efficiency as well as cost, because the instantaneous demand of PAA is dependent on the quality of the wastewater effluent being treated (i.e., organic matter, dissolved solids). Additionally, the impacts PAA may have on the wastewater effluent quality can be evaluated, particularly with respect to other discharge permit requirements (i.e., biochemical oxygen demand [BOD]). This approach also provides confidence in assessing feasibility and implementation logistics as the program moves from bench- to full-scale application.

As with other technologies, evaluation of PAA is often best conducted in a phased approach that incorporates bench-, pilot-, and full-scale testing before full-scale implementation (Figure 7.1). For example, Metro Vancouver, a federation of 21 municipalities in British Columbia, Canada, has used bench-scale testing to determine both the effective dose for microbial compliance as well as toxicity impacts at various PAA residuals. The information, developed during pilot testing, was used to inform the applied disinfection dose as well as the maximum allowable PAA residuals that are used in full-scale implementation.

In this approach, each phase of testing provides an increasing level of confidence in the development of design criteria for PAA as a wastewater disinfectant, minimizing performance risk as well as supporting a more accurate economic evaluation. Although not all testing phases are required to implement full-scale PAA disinfection, these steps can also inform the necessary background on regulatory and other site-specific requirements that need to be considered as part of a disinfection-alternatives analysis, design, and implementation program.

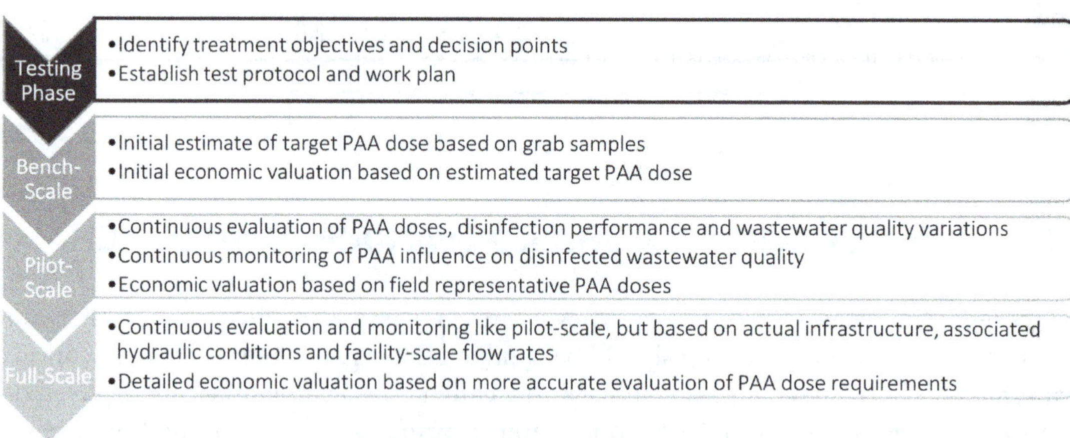

FIGURE 7.1 Peracetic acid disinfection evaluation as a phased approach.

It is important to establish the treatment objectives that serve as the basis for both treatment performance and cost comparisons to evaluate PAA as an alternative disinfection wastewater strategy. Typical discharge permits for publicly owned water resource recovery facilities (WRRFs) include requirements for bacterial indicators as well as other parameters that may be impacted by the practice of PAA disinfection. These parameters include BOD, carbonaceous BOD (CBOD), chemical oxygen demand (COD), total suspended solids (TSS), dissolved oxygen, pH, whole effluent toxicity (WET), and, for some receiving waterbodies, other key parameters such as disinfection byproducts (DBPs) that may include dichlorobromomethane, dibromochloromethane, cyanide, and/or dioxins.

1.1 Bacterial Compliance

Evaluation of bacterial inactivation is a critical part of PAA performance testing to prove disinfection efficacy. For decades, epidemiological studies have been used to evaluate how fecal indicator bacteria concentrations are associated with health effects of primary contact recreation. The U.S. Environmental Protection Agency's 1986 criteria were developed using epidemiological data collected in the late 1970s and early 1980s. In those studies, enterococci and *Escherichia coli* exhibited the strongest correlations to swimming-associated gastroenteritis. Both indicators continue to be used in epidemiological studies conducted throughout the world, including in the European Union and Canada (European Parliament & Council of the European Union, 2006).

The World Health Organization (WHO) recommends the use of enterococci as water quality indicators for recreational waters (WHO, 2003). Meta analyses and systematic reviews of epidemiological studies conducted worldwide show that these indicators generally provided substantial improvements over the indicators that were favored previously, such as total and fecal coliforms (Prüss, 1998; Wade, Pai, Eisenberg, & Colford, 2003; Zmirou, Pena, Ledrans, & Leterte, 2003).

Current recreational water quality criteria (RWQC) in the United States (U.S. Environmental Protection Agency, 2012) is a risk-based value for either enterococci or *E. coli* in fresh water and enterococci in marine water, although there are agencies that still use fecal and total coliform as indicators, as described further in Chapter 8. U.S. EPA has also reported that use of bacteriophage as a viral indicator could potentially serve to provide a higher level of protection of human health (U.S. EPA, 2015). Although it was not added to the criteria during the most recent review of water quality criteria, this indicator could be included at a later time as more data become available.

2.0 BENCH-SCALE TESTING

To establish planning-level PAA chemical and equipment requirements under a PAA disinfection alternative, a bench-scale test can be conducted to develop disinfection dose-response data to determine a preliminary PAA dose for application. The test involves collecting samples of an undisinfected secondary or tertiary wastewater and dosing a range of anticipated PAA concentrations. The PAA residual and bacterial concentrations are measured at various contact times following the addition of the chemical in stirred laboratory containers.

2.1 Development of an Experimental Protocol for Bench Testing

An experimental protocol should be developed to identify the potential range of PAA doses and water quality parameters to be evaluated at discrete contact times. This protocol should be developed considering the facility contact basin, if one already exists. For secondary effluent, typical doses of PAA can range from 0.5 to 3.5 mg/L to meet permit compliance, depending on undisinfected bacteria concentrations and instantaneous PAA demand of the site-specific effluent. Data from bench tests can be used to develop preliminary microbial dose-response curves and allow an assessment of whether water quality parameters of interest (e.g., CBOD, pH) may be impacted at the target PAA doses.

The experimental test plan should include a range of PAA doses, contact times that are reflective of the existing facility (for retrofits), PAA residual, and other parameters that are important to permit compliance. From an experimental standpoint, most typically, undisinfected secondary or tertiary effluent is collected and immediately treated and tested. It is critically important to measure an untreated sample as a control for each sample set to establish the initial bacteria concentration. Samples should be measured into sterilized beakers and placed onto a stir plate or jar testing device. In addition to testing disinfection efficacy at various doses, it may be of interest to also evaluate ultraviolet transmittance (UVT), particularly if the application being evaluated is to supplement an underperforming UV system. Because PAA is an oxidant, it can improve (increase) UVT and, in some cases, reduce TSS, which can improve the conditions for ultraviolet (UV) disinfection. Table 7.1 provides an example experimental matrix that encompasses the various water quality parameter measurements and laboratory analyses.

As shown in Table 7.1, data are collected and can be evaluated to determine PAA efficacy with respect to the permitted pathogens and effects on water quality parameters. The data from a single bench-scale test reflects

TABLE 7.1 Sample bench-scale experimental matrix, with results from Crooked Creek Water Reclamation Facility.

PAA dose (mg/L)	Contact time (minutes)	PAA residual (mg/L)	Fecal coliform (CFU/100mL)	pH	CBOD (mg/L)
Control	0	0	1400	7.3	<2.0
			1400		
0.5	10	0.18	500	7.4	<2.0
	30	0.14	86		
1.0	10	0.57	142	7.4	<2.0
	30	0.52	5		
2.0	10	1.37	11	7.4	<2.0
	30	1.35	0		
3.0	10	2.22	6	7.2	2.4
	30	2.02	0		

only a snapshot of the facility operation; if the test indicates that PAA doses resulted in water meeting permit requirements, additional bench-testing (reflective of a variety of operational conditions), or a pilot-scale or full-scale test may be considered. The dose-response curve developed from this type of a test can be used to provide a preliminary evaluation of PAA efficacy and to estimate pilot- and full-scale performance. Where the applied PAA doses do not meet permit requirements, consideration should be given to the possibility of repeating the test with a different range of PAA doses or of abandoning further PAA evaluation, depending on the drivers and cost-sensitivity for the specific facility.

Required log-reduction of microbial surrogates to meet permit limits is determined from microbial counts in the control sample. Disinfected sample results indicate the minimum required contact time to achieve that log inactivation. Collection of PAA residual immediately or very shortly after PAA is added (30 seconds to 1 minute) can aid in estimating the instantaneous PAA demand. The bacterial data collected at the various contact times can be used to identify the target dose and contact time needed to meet disinfection compliance. These results also serve as a starting point for subsequent testing phases.

2.1.1 Peracetic Acid Dose Calculation and Assay Confirmation

As previously described in Chapter 3, PAA decays rapidly when it is introduced to water and diluted from its equilibrium solution. Thus, when PAA is dosed into wastewater samples for jar tests, there are several important procedures for dosing and validation of the dose. In all cases, PAA should be dosed into samples as a neat solution, and not diluted before use. The chemical assay can be checked by dosing a known amount of PAA into a volume of distilled or deionized water that is being mixed and then immediately testing for PAA residual.

The PAA dose is calculated based on PAA in the stock solution (commercially available solution). The following equation can be used to determine the appropriate microliters of PAA that should be added to a 1-L sample in order to achieve a 2 mg/L dose, where V_{stock} is the volume of a 15% stock solution of PAA having a specific gravity of 1.16.

$$V_{stock}\,\mu L = \frac{\left(2\frac{mg}{L}\right)(1\,L)}{\left(0.15\frac{g\ of\ PAA}{g\ of\ solution}\right)\left(1.16\times 10^{3}\frac{mg}{mL}\right)}\times 10^{3}\frac{\mu L}{mL} \tag{7.1}$$

For a stock solution of a different strength, replace the 0.15 [for 15% strength] with the strength of the solution used and adjust the specific gravity, as needed.

2.1.2 Bench Test Procedures

The water quality parameters identified in the test protocol should be measured on each undisinfected sample using the appropriate method for each parameter. In the United States, National Pollutant Discharge Elimination System (NPDES) permits require either U.S. EPA or *Standard Methods* guidance for permit monitoring and compliance.

Jar tests should be conducted immediately upon collection and delivery to the laboratory. If it is not possible to analyze samples immediately, they can be stored on ice for up to 24 hours; however, initial bacteria concentrations may experience minor die off, although test results can still be used for technology screening and planning-level dose estimates. Three to four PAA doses and two to three contact times are generally appropriate for each of the tests, as shown in Table 7.1. Contact times should be selected based on anticipated minimum and average detention times, and possibly one more based on normal daily peak flow. For the jar test, the wastewater sample should be measured into aliquots and poured into cleaned and disinfected beakers on a jar tester apparatus, as shown in Figure 7.2.

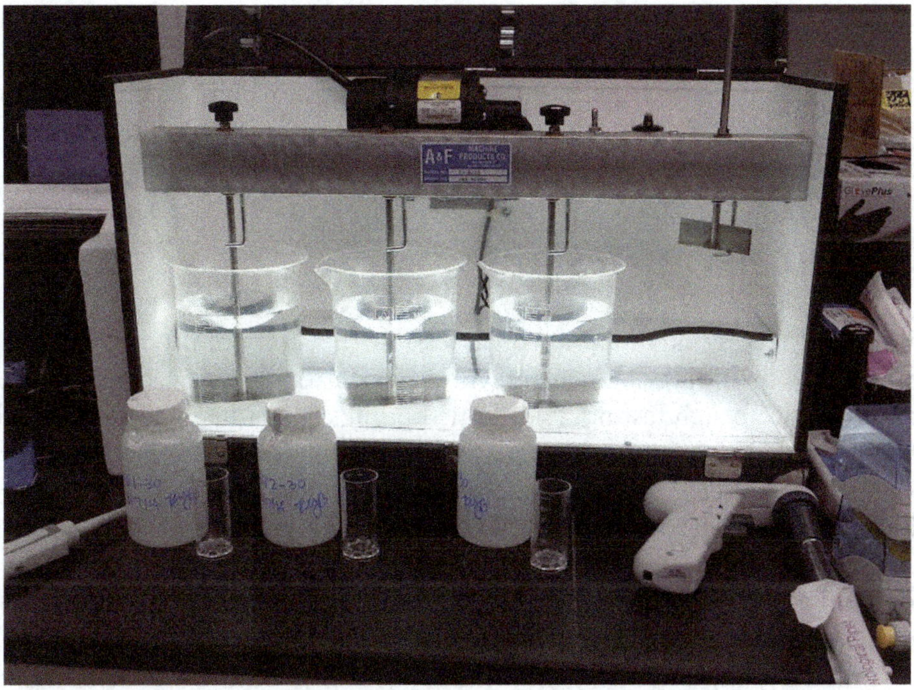

FIGURE 7.2 Typical jar testing apparatus used for PAA bench testing (Reprinted with permission from the International Ozone Association).

Stirrers should be set to provide continuous mixing, and PAA should be dosed into the stirred beakers using a micropipette at time zero (0 minutes). At the specified contact times defined in the test protocol, PAA residuals should be measured. Bacteria samples should be collected and must be quenched at each targeted contact time to stop the disinfection reaction from continuing. Quenching is typically conducted using excess sodium thiosulfate in the bacteria sample container, but sodium bisulfite can also be used at a dose of 1 g/L. After the PAA residual and bacteria samples are collected at the last contact time, the remainder of the water quality parameters (e.g., pH, dissolved oxygen, UVT) should be measured on the treated sample. Any samples that require specialized laboratory analysis should be collected and preserved according to the appropriate method and immediately packed on ice for delivery to the laboratory.

2.1.3 Bacterial Enumeration

Fecal indicator bacteria, which is used as the basis of monitoring in discharge permits, can be enumerated using various methods, including those in which the organisms are grown (cultured) and those in which their DNA is extracted from an environmental sample, amplified, and quantified (using quantitative polymerase chain reaction). These different methods result in method-specific units and values, as discussed in Chapter 1. Although there may be challenges with each of these enumeration techniques, the method selected for bench-scale testing should match what is used for permit monitoring and compliance.

3.0 PILOT-SCALE TESTING

Pilot-scale testing is valuable for assessing the site viability of a chosen technology, including PAA disinfection over a longer period than can be captured with bench testing. Pilot-scale findings can either lead to full-scale testing, implementation, or elimination of the technology from further consideration. The reasons for conducting pilot testing may vary because each WRRF is different. Some of the drivers for testing PAA are new permit requirements, including those for DBPs; cost, especially for facilities that use breakpoint chlorination; elimination of multiple chemicals for disinfection and quenching; and receiving water quality requirements.

Typically, the following minimum factors should be considered to advance PAA as an alternative disinfection method: demonstration of overall bacteria inactivation efficacy; development and understanding of the required dose-contact time relationship; and economics of the process. Additionally, testing should evaluate pH, DBPs, BOD or COD impacts, WET, and salts addition (compared to sodium hypochlorite, if total dissolved solids are of concern in the effluent).

3.1 Pilot Equipment and Site Selection

Selecting a location for a pilot test involves several considerations. Site selection should consider locations near the source of the undisinfected wastewater effluent, if possible, and the route needed to return the pilot effluent back to the facility process. When using mobile systems available through PAA suppliers, some of the site requirements will be identified by the technology provider. In general, these include: power requirements, potable water needs (for safety and wash-down), wastewater and chemical spill containment, and the minimum site footprint. Additional attention may be needed for equipment delivery access, placement of the unit, and whether a forklift or other equipment is necessary to locate the pilot system in the selected space. Figure 7.3 illustrates a typical mobile pilot reactor assembly, including the chemical feed and containment system (shown at far right).

Additional site preparation may be needed in locations with freezing potential, so equipment can be winterized, if necessary. During cold weather, small-diameter piping, such as sample taps, are prone to freezing. Simple

FIGURE 7.3 Mobile PAA pilot system, sited at the Metro Wastewater Reclamation District Robert W. Hite Treatment Facility in Denver, Colorado (Photo courtesy of Edyta Stec-Uddin).

tarps weighted with concrete blocks can work well if properly installed for windy conditions; however, it is also important to allow access to the sample ports, instrumentation, and other parts of the pilot system for sampling and monitoring. This type of protection may not be sufficient to prevent sampling points from freezing, especially if sample ports are small-diameter and/or trap water when not in use. Use of heat tracing or pipe insulation along the small-diameter pipe insulation or an outdoor heater may be necessary for mobile pilot equipment placed outdoors, if conditions are extremely cold.

Spill management controls are also necessary. While spill prevention is the first line of defense against a chemical release, planning for the unexpected wastewater or chemical spill is essential, and any potential spills must be contained or redirected to an approved location. Reliable flow is necessary to evaluate disinfection efficiency; continuous flow through the pilot unit is important to ensure that the system is providing a consistent dose and contact time for representative samples. Selecting a location for the influent pump can be a challenge if there are issues with algae growth or variable water levels in the trough where the feed pump is installed. To avoid these issues, make sure the influent pump suction is elevated away from the floor of the feed structure, but still deep enough to provide adequate flow, even during low-flow conditions.

3.2 Development of Experimental Plan for Pilot Testing

Each site is unique and may or may not have the means to perform laboratory testing on-site. Locations with a laboratory may need extra personnel to perform additional testing. Testing requirements, frequency, and pilot duration should be clearly specified in sampling and testing protocols and communicated with local operations and laboratory staff. Table 7.2 shows a sample pilot test matrix. The test plan would be repeated for each specific test condition being evaluated (i.e., dose and contact time).

Sampling protocols should be realistic and provide as much detail as possible, including the number and frequency of tests. Some pilot units allow testing from different sample ports, thus allowing conditions to be tested at various detention times. Information should be included in the sampling protocol regarding ports to be tested. If an on-site laboratory is not available, it is even more important to provide a detailed sampling schedule and agree on conditions with the external laboratory, including sample date, time, number of samples, and needed analysis. Before piloting, draft test plans can be shared with state and federal regulatory agencies to get feedback that can be incorporated to the final plan.

3.2.1 Safety

As with other chemicals, safe storage, handing, and feed systems are important for PAA. Typically, the pilot provider will arrange safety training, review the related hazards, and explain spill response as well as response to accidental contact with exposed skin and other concerns. An eye wash, either permanent or stand-alone, must be available near the PAA storage container and dosing pump. If the pilot is scheduled during cold weather conditions, the eye wash may need to be winterized. An emergency contact list (laminated to avoid weathering) with phone numbers should be posted in a visible place; this list is in addition to any other safety requirements, including posting the safety data sheet (SDS) for the product.

3.2.2 Daily Checks

The pilot should be checked daily, at a minimum, to be sure that it is in working condition, it hasn't developed leaks, and it does not have anything impeding the flow through the unit. The PAA dosing pump and the wastewater feed pump should also be checked at this time, including inspection of pilot tubing and hoses. If the pilot unit is equipped with a flow meter, the meter reading should be included in daily checks because abnormal values can indicate problems with flows.

3.2.3 Online Instrumentation

In addition to grab samples, online instrumentation may provide valuable information. Because side-stream pilot flows are generally small, an intermediate tank between the pilot outlet and the final discharge point may be

TABLE 7.2 Sample experimental matrix for pilot-scale testing.

Water quality parameter	Sampling location: Influent (upstream of PAA feed)	Sample Port A	Sample Port B	Sample Port C	Frequency	Sample type
E. coli (MPN[a]/100 mL)	✓	—	✓	✓	2X/day	Grab
PAA residual	—	✓	✓	✓	2X/day	Grab
TSS (mg/L)	✓	—	—	✓	Continuous	Online
BOD_5 (mg/L)	✓	—	—	✓	Weekly	Grab
CBOD (mg/L)	✓	—	—	✓	Weekly	Grab
pH	✓	—	—	✓	Continuous	Online
Water temperature	✓	—	—	—	Continuous	Online
WET	—	—	—	✓	Once during pilot testing	In accordance with permit
Trihalomethanes (µg/L)	—	—	—	✓	Once during pilot testing	Grab
Haloacetic acids (µg/L)	✓	—	—	✓	Once during pilot testing	Grab
Cyanide (µg/L)	✓	—	—	✓	Once during pilot testing	Grab
N-Nitrosodimethylamine (ng/L)	✓	—	—	✓	Once during pilot testing	Grab
Sodium, sulfate, and chloride (mg/L)	✓	—	—	✓	Once during pilot testing	Grab
Dissolved oxygen (mg/L)	✓	—	—	✓	Continuous	Online

[a]Most Probable Number

needed to capture an adequate volume of sample. During the pilot study at the Metropolitan Wastewater Reclamation District (MWRD) Robert W. Hite Treatment Facility, a tote was used for this purpose. Pilot effluent was redirected to the tote with several online instruments measuring dissolved oxygen (DO), TSS, temperature, and pH. During the study, the tote was painted black using spray paint to prevent sunlight interference with optic probes. Figure 7.4 illustrates the installation, including secondary containment for spill prevention.

During the pilot study at the MWRD Robert W. Hite Treatment Facility, algae growth was observed on the influent strainer (Figure 7.5), requiring daily cleaning. A second strainer was purchased to minimize the time the pilot unit was out of service when replacing the strainer.

FIGURE 7.4 Installation of tote to house in-line instrumentation for continuous monitoring (Photo courtesy of Edyta Stec-Uddin).

Biofilm formation was also observed during the pilot study (Figure 7.5). To address the issues associated with this biofilm, a "flush" operation (Figure 7.6) was implemented twice a week, which involved significantly raising the speed of the influent pump for about 10 to 15 minutes until no biofilm was visible. This maintenance operation was coordinated with the sampling schedule so that no samples were collected immediately following the flush.

FIGURE 7.5 Algae growth and biofilm on the pilot influent strainer (Photo courtesy of Edyta Stec-Uddin).

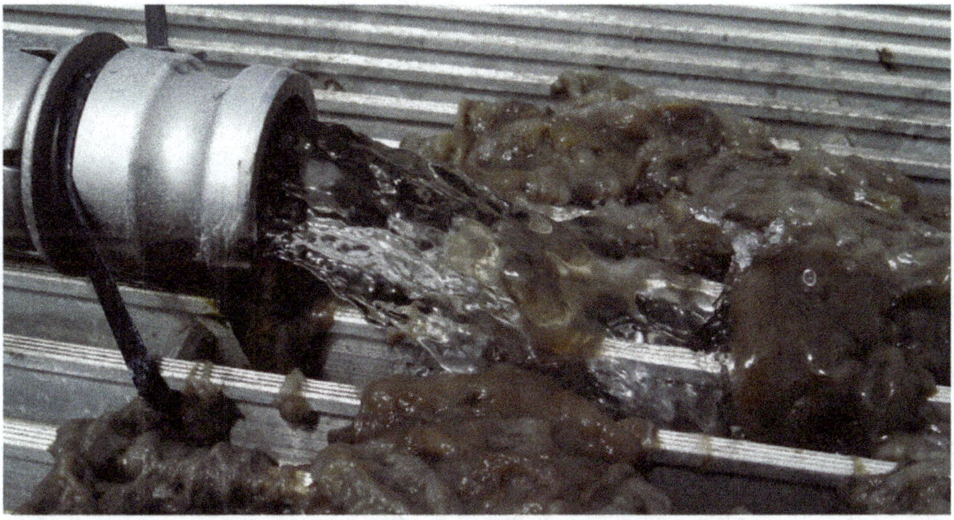

FIGURE 7.6 Biofilm released from the MWRD pilot during maintenance flush (Photo courtesy of Edyta Stec-Uddin).

4.0 FULL-SCALE TESTING

Full-scale testing provides the ultimate proof-of-concept for a specific treatment facility and water matrix. Full-scale testing can be beneficial in evaluating a range of test conditions that are not reflected in a single grab sample used for bench testing. Diurnal changes in effluent quality and flow and their effect on PAA disinfection can be observed and quantified. Seasonal variations can be evaluated if full-scale testing is conducted over a long enough period.

The ideal configuration is to compare PAA disinfection with a facility's existing disinfection system in a side-by-side trial. This approach was used at the F. Wayne Hill Water Resources Center (FWH WRC) in Gwinnett County, Georgia, when PAA was used as a temporary back-up to the facility's ozone disinfection system during a construction project. After completing jar tests to determine the PAA dose required to meet disinfection targets, a full-scale test plan was developed for full-scale implementation. The sample plan included key parameters for disinfection, as well as pertinent FWH WRC NPDES parameters, including pH and COD.

At FWH WRC, ozone is used for final disinfection before discharge into Lake Lanier in a planned potable reuse scenario. There are four ozone contactors operating in parallel for final disinfection, so PAA could be applied to a single contactor while maintaining ozone disinfection at the other contactors in service. Full-scale tests were conducted in November 2014 through January 2015; at that time, only two contactors were in service because the facility flow was running at approximately one half of the design capacity. Ozone was applied in one contactor, and PAA was applied in the other contactor. Figure 7.7 shows the configuration of the contactors and the parameters monitored at two sampling locations following PAA or ozone application.

The temporary PAA feed system, supplied by PeroxyChem, included a 15% solution PAA chemical tote placed on a plastic spill containment structure, a self-contained weatherproof pump skid with programmable logic control (PLC), and PAA feed piping. Figure 7.8 provides a photograph of the general arrangement of the system located beside the inlet feed channel to the post-ozone contactors. Peracetic acid was applied as neat solution using an electrically driven peristaltic pump (one duty, one standby); PAA feed was flow-paced by the PLC, which received a flow signal from the facility's supervisory control and data acquisition (SCADA) system. The feed system was equipped with appropriately placed pressure relief valves with outlets routed to containment; PAA feed piping was constructed of passivated stainless-steel tubing, compatible with PAA. Piping was installed to drip PAA onto the surface of the inlet channel to maximize hydraulic mixing of PAA into the main flow.

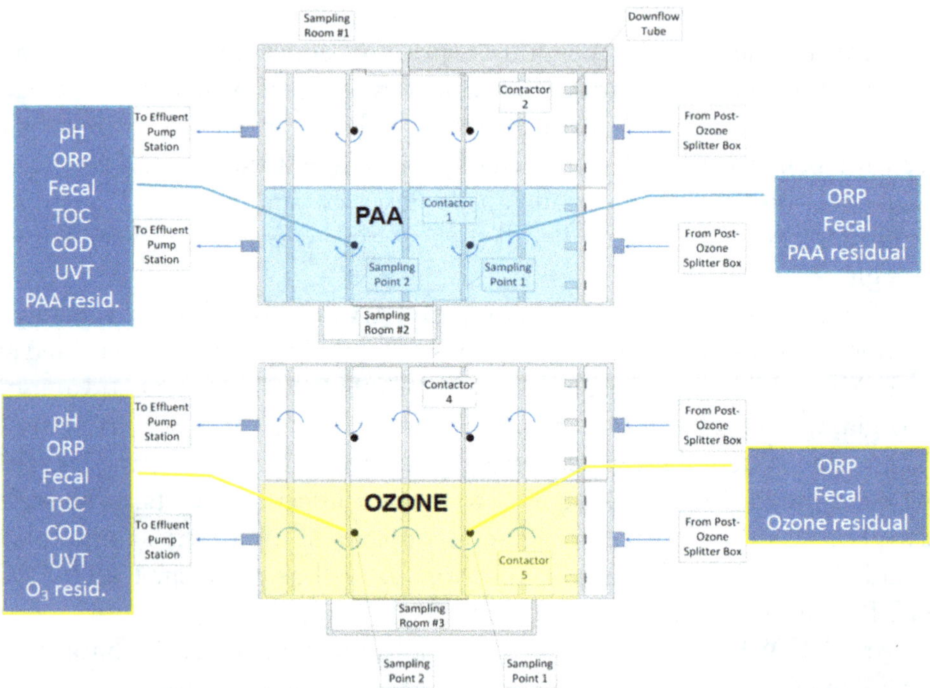

FIGURE 7.7 Schematic of full-scale, side-by-side disinfection trial for ozone and PAA at FWH WRC (Reprinted with permission from the International Ozone Association).

FIGURE 7.8 The FWH WRC temporary full-scale PAA disinfection feed system (Photo courtesy of Denise Funk).

Jar tests that were conducted before installation of the full-scale pilot system provided a benchmark for target PAA doses. During full-scale testing, three different PAA doses (1.5 mg/L, 1 mg/L, and 0.75 mg/L) were each tested during 6-hour test periods using three rounds of sampling. Results from this trial indicated that a PAA dose of 1.5 mg/L achieved disinfection to support compliance with the NPDES discharge permit.

Full-scale testing was also performed at the Crooked Creek Water Reclamation Facility in Gwinnett County, Georgia. The pilot at this site was installed in December 2014 as a stand-alone disinfection system to serve as a supplement or provide backup to an existing medium-pressure UV disinfection system that was experiencing performance challenges. The PAA was dosed immediately after tertiary filtration before entering the existing UV system. PeroxyChem supplied the temporary PAA feed system, enclosed in weatherproof containers; the skid included two (one duty, one standby) peristaltic pumps with a flow-paced control system. Figure 7.9 shows the PAA feed and sample locations at the Crooked Creek Water Reclamation Facility.

Combined tertiary effluent flows at the Crooked Creek Water Reclamation Facility enter a UV influent channel that flows into four separate UV reactors. Following UV disinfection, disinfected effluent flows to the post-aeration basin before discharge.

Before full-scale implementation, bench-scale dose-response testing was performed to determine a target PAA dose for disinfection. Data from bench

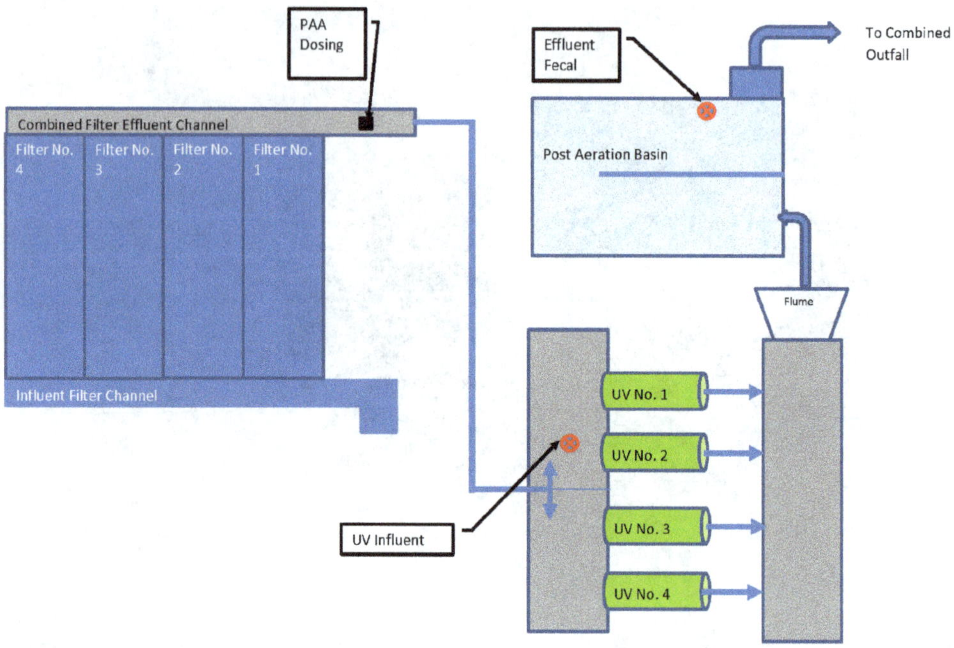

FIGURE 7.9 Crooked Creek Water Reclamation Facility temporary full-scale PAA disinfection schematic (red circles indicate the sample locations used during the pilot trial) (Graphic courtesy of Denise Funk).

tests were used to develop dose-response curves as well as support an initial evaluation on the impact of PAA on water quality, including pH and 5-day CBOD. The full-scale pilot, during trials in April 2015, confirmed bench-scale test results that showed PAA could meet bacteria limits with a dose of 1 mg/L at average flow. During the following summer, higher PAA doses (up to 3 mg/L) were necessary to meet disinfection criteria; this was likely attributable to fluctuations in the upstream treatment processes.

5.0 DEFINING DESIGN DOSE

The residual concentration times contact time (C_RT) approach typically used for chlorine disinfection is not fully adequate to assess the effectiveness of PAA disinfection because of the rapid decay of PAA residual and deviations from first-order kinetics. There are several microbial inactivation models that can be used to account for the decay of PAA over time; these models are described in Chapter 3. Because these inactivation models are empirical, they do not reflect a mechanistic description of PAA disinfection but are instead specific to the wastewater characteristics at each application.

5.1 Data Evaluation

Peracetic acid disinfection depends on a combination of factors, including contact time and residual concentrations. To establish site-specific PAA dose requirements, experimental inactivation data combined with residual PAA measurements at various contact times can be fit using any of the appropriate inactivation models by least squares regression analysis. This form of kinetic modeling supports accurate design dose estimates, which are critical for accurately assessing the costs of operation and conducting an economic evaluation of PAA as a disinfection alternative.

The models that can be used include a modified Chick-Watson, incomplete gamma function, Hom's model, Power Law, or the Hom–Power Law (Dunkin et al., 2017; Santoro et al., 2007), as described in Chapter 3. Any of these kinetic models can be fit to the site-specific data, assuming the contact time can be estimated. Inactivation curves are then prepared using the parameters determined from the model fitting for various contact times. Figure 7.10 presents an example of the inactivation curves developed at the Crooked Creek Water Reclamation Facility from bench-scale testing.

Contact times at an existing facility, where PAA may be retrofit to the existing infrastructure, will likely be predetermined based on the site. Although a new PAA disinfection system can be designed around a desired contact time, retrofitting existing infrastructure will determine the available contact times at the treatment facility. It is also important to consider the contact times during peak hour and average flow.

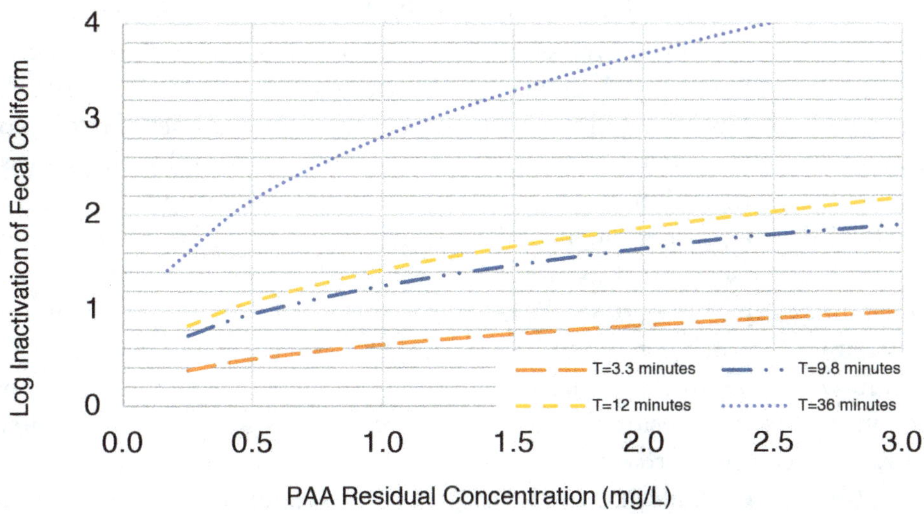

FIGURE 7.10 Sample graph of predictive microorganism inactivation fit to Hom's model from the Crooked Creek Water Reclamation Facility bench-scale testing data (Chart courtesy of Amy Gruss).

5.2 Development of Design Dose

As described, PAA demand and decomposition kinetics are unique for specific wastewaters and must be characterized for each specific site (Haas et al., 1996); this can be conducted at bench scale by testing wastewater collected from multiple time periods. Additionally, an enhanced understanding of the process can be developed through pilot-scale testing. Proof of performance is demonstrated by full-scale piloting, where predicted inactivation curves can be used to target initial doses that can be optimized at full scale; this accounts for differences in contactor hydraulics, water quality variability, and other factors that may not be adequately represented in bench- or pilot-scale testing.

Permit compliance includes requirements for bacterial inactivation. However, other discharge parameters must also be considered, including pH, CBOD, WET, and other parameters. Though kinetic models can help predict a required dose under normal wastewater conditions, it is important to incorporate safety factors to address potential variability in bacterial concentrations after secondary treatment because of changes in weather or wastewater facility operations. And, while an increased PAA dose may effectively inactivate microorganisms, the increased dose also has the potential to affect these other discharge parameters; thus, it is important to test and assess higher doses than are estimated for meeting disinfection compliance to ensure other permit requirements are still met under a "high demand" scenario.

The modeled relationship between contact times and residual under varying flowrates provides a theoretical PAA dosing range for minimum, average, and maximum disinfectant demand conditions. These doses incorporate the immediate PAA demand estimate from bench- and pilot-scale tests; however, depending on the contactor configuration, a safety factor may be needed to address hydraulic and other influences on the full-scale process.

Finally, while Chapter 9 provides more detailed descriptions of PAA disinfection process control, most dosing schemes are based on flow pacing. This type of dose control assumes that the wastewater characteristics do not significantly change or, if they do, the dosing scheme will be manually adjusted within the controlling system. Other dosing schemes are based on both flow and parameters, such as secondary effluent color or turbidity, or the amount of PAA residual in the disinfected wastewater, and are applied either in a feed-forward or feedback loop. However, online, real-time sensors, although available, are often labor intensive, requiring almost constant attention. As more research is conducted to identify better online proxies or surrogates for PAA demand and residual, it is anticipated that more sophisticated control strategies will be used.

Currently, however, development of data through bench-, pilot-, and full-scale trials are the best opportunity to identify PAA doses for implementation. Leveraging basic process control methods (e.g., flow pacing), facilities can work to optimize a dosing scheme once a better understanding of PAA disinfection performance is developed through full-scale application.

6.0 REFERENCES

European Parliament & Council of the European Union. (2006). Directive 2006/7/EC of the European Parliament and of the Council of 15 February 2006 Concerning the Management of Bathing Water Quality and Repealing Directive 76/160/EEC. *Official Journal of the European Union,* L64, 31–51. Retrieved from http://europa.eu.int/eurlex/lex/LexUriServ/site/en/oj/2006/l_064/l_06420060304en00370051.pdf

Dunkin, N., Weng, S., Schwab, K. J., McQuarri, J., Bell, K., & Jacangelo, J. G. (2017). Comparative inactivation of murine norovirus and MS2 bacteriophage by peracetic acid and monochloramine in municipal secondary wastewater effluent. *Environmental Science & Technology,* 51, 2972–2981.

Haas, C. N., Joffe, J., Jacangelo, J. G., Anmangandla, U., & Heath, M. (1996). Water quality and disinfection kinetics. *American Water Works Association Journal, 88*(3), 95.

Prüss, A. (1998). Review of epidemiological studies on health effects from exposure to recreational water. *International Journey of Epidemiology, 27*(1), 1–9.

Santoro, D., Gehr, R., Bartrand, T. A., Liberti, L., Notarnicola, M., Dell'Erba, A., Falsanisi, D., & Haas, C. N. (2007). Wastewater disinfection by peracetic acid: Assessment of models for tracking residual measurements and inactivation. *Water Environment Research,* 79(7), 775–787.

U.S. Environmental Protection Agency. (2012). *Recreational water quality criteria* (EPA-820/F-12-058). Washington, DC: U.S. Environmental Protection Agency, Office of Water.

U.S. Environmental Protection Agency. (2015). *Review of coliphages as possible indicators of fecal contamination for ambient water quality* (EPA-820/R-15-098). Washington, DC: U.S. Environmental Protection Agency, Office of Water.

Wade, T. J., Pai, N., Eisenberg, J. N. S., & Colford, J. M., Jr. (2003). Do U.S. Environmental Protection Agency water quality guidelines for recreational waters prevent gastrointestinal illness? A systematic review and meta-analysis. *Environmental Health Perspectives, 111*(8), 1102–1109.

World Health Organization. (2003). *Guidelines for safe recreational water environments: Volume 1, Coastal and fresh waters*. Retrieved from http://www.who.int/water_sanitation_health/bathing/srwe1/en/

Zmirou, N., Pena, L., Ledrans, M., & Leterte, A. (2003). Risks associated with the microbiological quality of bodies of fresh and marine water used for recreational purposes: Summary estimates based on published epidemiological studies. *Archives of Environmental Health: An International Journal, 58*(11), 703–711.

7.0 SUGGESTED READING

Ferguson, D. M., Griffith, J. F., McGee, C. D., Stephen, B., Weisberg, S. B., & Hagedorn, C. (2013). Comparison of Enterococcus species diversity in marine water and wastewater using Enterolert and EPA Method 1600. *Journal of Environmental and Public Health,* 1–6.

Sercu, B., van de Werfhorst, L. C., Murray, J. L. S., & Holden, P. A. (2011). Cultivation-independent analysis of bacteria in IDEXX. *Applied and Environmental Microbiology*, 77(2), 627–633.

8

Regulatory Coordination

Kati Bell & George Garde

1.0 INTRODUCTION

In the United States and Canada, wastewater effluent discharges are regulated federally, and the requirements are issued by local (by state in the United States and provincially in Canada) regulatory authorities. As part of these regulatory frameworks, disinfection is typically required to meet discharge requirements. Peracetic acid (PAA) is a chemical that can be used to meet these discharge requirements for bacterial compliance. While bacteria compliance is the primarily aim of PAA addition, the treatment process is also subject to other regulatory requirements, including PAA residual. While PAA residual is the most obvious secondary permit requirement, PAA can also have effects on other normal discharge water quality parameters, such as pH or dissolved oxygen.

In 2015, there were also significant regulatory changes impacting chlorine disinfection, which is still the most commonly used disinfection technology in North America (Bell & Wylie, 2016). In the United States, the U.S. Environmental Protection Agency (U.S. EPA) updated its ambient water quality criteria for human health. This update provided relaxation for concentrations of trihalomethanes; however, criteria for cyanide were lowered by nearly an order of magnitude, from 140 to 4 μg/L, for "Consumption of Water + Organism" (U.S. EPA, 2015).

Cyanide commonly occurs in municipal wastewaters, and a growing number of water resource recovery facilities (WRRFs) across the United States are detecting cyanide in chlorinated effluents at levels exceeding influent concentrations (Deeb et al., 2003). This is particularly important as regulatory requirements for nutrient removal are being implemented across the country; related changes in effluent characteristics may increase cyanide formation potential, which has been the subject of several studies (Pandit et al., 2006; Zheng et al., 2004). Additionally, it is anticipated that U.S. EPA will eventually revisit other disinfection products such as nitrosamines and dioxins because the agency did not update human health criteria for these chemicals as a result of outstanding technical issues at the time of the criteria update (U.S. EPA, 2015).

In Canada, Environment Canada began work on a national strategy to manage wastewater effluents under the direction of the Canadian Council of Ministers of the Environment (CCME) in 2006. In 2007, CCME released the *Draft Canada-wide Strategy for the Management of Municipal Wastewater Effluent*. At the same time, Environment Canada published the *Proposed Regulatory Framework for Wastewater* to explain how the Canada-wide strategy would be implemented. A draft regulation based on the CCME strategy was released for comment in 2010 and, on July 28, 2012, the Wastewater Systems Effluent Regulations were published in the *Canada Gazette*. The regulations, which went into effect in January 2015, are made under the Fisheries Act, which prohibits unauthorized deterioration, disruption, and destruction of fish habitat. The new regulations are prompting many Canadian facilities that have historically used chlorination to consider how to both avoid outstanding issues related to chlorine as well as maintain regulatory compliance.

1.1 Water Quality Criteria Development in the United States

In the United States, the foundation of any wastewater effluent discharge permit is the Clean Water Act (CWA), which originated from the Federal Water Pollution Control Act of 1948. Subsequent amendments to the law officially re-established the statute as CWA in 1972 (U.S. EPA, 2015). Since then, multiple amendments have been made to progressively update CWA (i.e., 1977, 1981, and 1987). The CWA establishes requirements for water quality standards for surface waters and regulates the discharge of pollutants into U.S. waters, with permit limits based on technology, economics, and water quality. U.S. EPA rules, promulgated under authority of CWA, provide a minimum level of treatment for all wastewaters (with some rare variances). Wastewater disinfection is required where receiving waterbodies are designated for uses that have ambient water quality criteria for bacteria. Once a facility is permitted, when treated effluent meets these standards it may be discharged to a receiving waterbody that could ultimately be used for various purposes.

To protect the various designated uses of a waterbody, including those that serve as waters for contact recreation and drinking water supplies, CWA requires U.S. EPA to develop water quality criteria that accurately reflect the latest scientific knowledge. These criteria are based solely on data and scientific judgments of pollutant concentrations and environmental or human health effects, and are developed for the protection of both aquatic life and human health. There are generally two types of water quality criteria that are applicable when considering permitting for PAA disinfection: numeric water quality criteria and narrative water quality criteria.

1.1.1 Numeric Water Quality Criteria

Criteria developed under Section 304(a)(1) of the CWA serve as guiding recommendations for states and authorized tribes in their development of state or local water quality standards, under Section 303(c). When state or local water quality criteria are developed under 40 C.F.R. 131.11(b), states or authorized tribes should do the following:

- Adopt U.S. EPA's 304(a) recommendations;
- Adopt 304(a) criteria, but modify them based on site characteristics; or
- Develop their own scientifically based criteria (U.S. EPA, 2000).

Currently, U.S. EPA (2015) has published numeric water quality criteria for 150 pollutants to protect human health and aquatic life. There are microbial and chemical criteria that exist that are relevant to National Pollutant Discharge Elimination System (NPDES) permitting.

1.1.1.1 Bacteria

With respect to microorganisms (pathogens), water quality criteria are set to protect the public from exposure to harmful levels of pathogens during recreational activities such as swimming, or through drinking or eating fish/seafood from contaminated surface water. The extent of regulatory authority of the CWA includes protection of surface water as a supply source for drinking water as well as recreational and aquatic food source uses. U.S. EPA currently uses fecal indicator bacteria, *Escherichia coli* and enterococci, to determine the level of fecal contamination present in environmental waters, as described in Chapter 1. However, future recommendations may also include use of coliphages as possible viral indicators of fecal contamination in ambient water. Of note, there has been recent work on PAA showing that it is as effective as other mature disinfection technologies for inactivation of viruses in wastewater (Weng et al., 2018).

While U.S. EPA recommends *E. coli* and enterococci as the monitoring and compliance indicator organisms, some states have not yet fully implemented these criteria in their NPDES permits, and some states still enforce fecal coliform limits. Although some state agencies have claimed that the transition from fecal coliform to the new criteria will have a no-cost impact on WRRFs, there are fundamental reasons supporting otherwise (Bell & Wylie, 2016), particularly as it relates to PAA disinfection. This is because coliforms, fecal coliform, and *E. coli* are Gram-negative bacteria. Enterococci, on the other hand, are Gram-positive cocci that can persist in harsh environmental conditions (Švec & Devriese, 2009). Characteristic differences between Gram-positive and Gram-negative bacteria (i.e., Gram-negative bacteria do not have the thick mesh-like cell wall made of peptidoglycan that Gram-positive bacteria have) result in Gram-positive bacteria having higher resistance to disinfection.

1.1.1.2 Chemical Criteria

With respect to chemicals, water quality criteria, typically adopted by states and approved by U.S. EPA, are numeric values limiting the amount of chemicals that may be present in water to protect human health and aquatic life. Criteria are based on U.S. EPA guidance, as explained in the previous section. Such criteria are developed by assessing the relationship between pollutants and their effect on human and aquatic life, and they are used by states and tribes to establish water quality standards to ultimately provide a basis for controlling discharges or releases of pollutants to such waters.

Human health criteria are generally developed to provide guidance on the highest concentration of a pollutant in water not expected to pose a significant risk. However, there are conditions where discharges of chemicals that do not pose a direct risk may react with other chemicals in the water to form chemicals such as disinfection byproducts that do pose a risk. In 2015, U.S. EPA issued updated national recommended water quality criteria (RWQC), which contain updates for 94 chemical pollutants and can be accessed at http://water.epa.gov/scitech/swguidance/standards/criteria/health/.

While not all criteria are directly derived using human health information, recommended aquatic life criteria (ALC) developed under Section 304(a) of CWA reflect the "latest scientific knowledge" concerning "all identifiable effects" of pollutants in question. Since the early 1980s, U.S. EPA has been developing ALC to protect aquatic organisms from chemical-specific pollutants under Section 304(a) of CWA. In 1985, U.S. EPA published *Guidelines for Deriving Numerical National Water Quality Criteria for the Protection of Aquatic Organisms and Their Uses* (Stephan et al., 1985). The guidelines have provided uniformity and transparency in the derivation methodology of ALC for compounds for several classes of chemicals.

1.1.2 Narrative Water Quality Criteria

In accordance with 40 C.F.R. 131.11(b)(2), in adopting water quality criteria, states and authorized tribes should "establish narrative criteria or criteria based on biomonitoring methods where numeric criteria cannot be established or to supplement numeric criteria". Narrative criteria for pollutants with toxic effects can be established in state and authorized tribal water quality standards (WQS) in various forms; for example, a narrative toxic pollutant criterion could take the following (or similar) form: "Waters shall be free from toxic pollutants in toxic amounts".

Such narrative criteria serve as the basis for establishing pollutant or chemical-specific water quality based effluent limits (WQBELs) for NPDES discharges where the state or authorized tribe has not adopted chemical-specific numeric criteria for a pollutant. They can also serve as a basis for establishing whole effluent toxicity controls per 40 C.F.R. 122.44(d).

1.2 Canadian Water Quality Regulations

In Canada, a similar regulatory framework exists as in the United States. Environment Canada has a national strategy to manage wastewater effluents under the direction of CCME. In 2007, CCME released the *Draft Canada-wide Strategy for the Management of Municipal Wastewater Effluent* concurrently with *Proposed Regulatory Framework for Wastewater*, published by Environment Canada to lay out an implementation strategy.

The *Wastewater Systems Effluent Regulations* were finalized on July 28, 2012, and went into effect in January 2015.

The new regulations are the first federal regulations that specifically address municipal WRRF effluents and impose strict limits for final effluent quality, which had not been previously regulated by the provincial authorities, related to un-ionized ammonia, acute lethality testing, and total residual chlorine. Thus, like WRRFs in the United States, many Canadian facilities that have historically used chlorination are faced with challenges in meeting regulatory requirements.

The development of CCME's environmental quality guidelines for water, tissue, sediment, and soil follows a clearly defined, transparent process. This process is designed to enhance communication between the water and soil quality task groups, senior CCME committees, jurisdictions, stakeholders, and the public. In doing so, it reinforces overall CCME priorities, contributes to budget and work planning, allows for a more predictable production schedule, tracks progress, and enables public and stakeholder awareness and participation. The process ideally takes place over approximately 2 years, beginning with priority setting and ending with publication of fact sheets in a Canadian Environmental Quality Guidelines update. Technical and policy challenges do arise during guideline development that may require an extension of the timeline. The general process for developing guidelines includes the following steps:

1. Priority setting
2. Confirmation of task group work plans
3. Public notification
4. Guideline development
5. Peer and task group review
6. Public review
7. Final report and fact sheet
8. Translation of fact sheet
9. Environmental Planning and Protection Committee and Management Committee approval
10. Publication in Canadian Environmental Quality Guidelines Update

2.0 NATIONAL POLLUTANT DISCHARGE ELIMINATION SYSTEM PROGRAM

With respect to protecting the nation's waters for all designated uses, NPDES is a permit program under Section 402 of CWA that limits water pollution by regulating discharges from point sources (U.S. EPA, 2015). Point sources include industrial, municipal, or other facilities that discharge effluent (wastewater) or stormwater directly or indirectly into receiving surface waters. A subset of dischargers are the publicly owned treatment works that discharge treated wastewater from community wastewater collection systems. The NPDES permits are issued by authorized states, with the exception of four states (i.e., Massachusetts, New Hampshire, New Mexico, and Idaho) in which U.S. EPA is the permitting authority. There are two types of permits under the NPDES program: individual and general permits. Individual permits are issued for specific facilities, whereas general permits cover discharges from multiple facilities that are similar in nature.

2.1 Establishing Limits in a National Pollutant Discharge Elimination System Permit

The limits in a NPDES permit are established using two basic approaches for protecting and restoring the nation's waters. One is a technology-based approach, whereby the permitting authority sets permit conditions either based on national effluent limit guidelines and/or new source performance standards (i.e., minimum requirements that apply to all dischargers in a class, such as municipal WRRFs), or case-by-case on the permit writer's best professional judgment (CWA section 301(b) and 40 C.F.R. 125.3). The other approach includes WQBELs and is designed to ensure compliance with the water quality standards applicable to a waterbody. Where technology-based effluent limits would still leave a reasonable potential that discharge could cause or contribute to an exceedance of WQS, a more stringent WQBEL is imposed to achieve those WQS.

2.1.1 Bacteria

In the United States, NPDES limits for microbial indicators are most typically enforced at the "end-of-pipe", meaning that the ambient water quality criteria must be met at the end of the treatment process before it is discharged to the receiving waterbody. This issue is somewhat murky because, while the U.S. EPA, in documents such as the Ephraim King letter (U.S. EPA, 2008), has indicated that there is a prohibition on the use of mixing zones for bacteria in primary contact recreation waters, individual primacy states (i.e., states that have EPA granted authority) may, in fact, use mixing zones to calculate the effluent limits for bacteria. However, while the mixing zone calculation is allowable, most states typically implement bacteria criteria at the end-of-pipe and use the criteria directly in NPDES permits as discharge limits. Thus, in most discharge permits in the United States, bacteria limits are set equal to the RWQC for bacteria if the receiving waterbody is designated for a use that includes contact recreation.

2.1.2 Chemicals

If a waterbody is already meeting its water quality standards, then those standards are used in calculating the WQBEL for the NPDES permit by using a dilution factor approach that considers the low flow condition of the receiving stream or impoundment and the discharge flow contributed by the permittee. If a state adopts the U.S. EPA-recommended criteria as the basis of its water quality criteria, then there is the potential for the permit to include the criteria as the permit limit if the low flow in the stream is low or zero.

2.1.3 Whole Effluent Toxicity Testing

The whole effluent approach to toxics control for the protection of aquatic life involves the use of acute and chronic toxicity tests to measure the toxicity of wastewaters. Whole effluent toxicity (WET) is a useful parameter for assessing and protecting against impacts on water quality and designated uses caused by the aggregate toxic effect of the discharge of pollutants (Wang, 1990). Whole effluent toxicity tests use standardized, surrogate freshwater or marine (depending on the mixture of effluent and receiving water) plants, invertebrates, and vertebrates. U.S. EPA has published protocols listing numerous marine and freshwater species for toxicity testing (U.S. EPA, 1991a, 1991b). Chapter 5 provides additional information on WET testing protocols. Generally, WET testing addresses acute and chronic toxicity, where *acute* is defined as a test of 96 hours or less, in which lethality is the measured endpoint. A *chronic toxicity test* is defined as a long-term test in which sublethal effects, such as fertilization, growth, and reproduction, are typically measured, in addition to lethality.

2.2 Peracetic Acid Disinfection Permitting in the United States

The information required for obtaining a permit for PAA disinfection varies widely among regulatory agencies, depending on the state water quality calculation approaches as well as the size of the receiving waterbody, which is indicative of the dilution factor. A survey of U.S. NPDES coordinators was conducted in 2018 and, based on that survey, there are currently three general methodologies that have been applied for issuance of NPDES permits for PAA disinfection at the time of development of this manual. It should be noted that there is a theme of a 1-mg/L discharge residual limit that is observed as a baseline for regulatory determinations. The 1.0-mg/L criterion is derived from the U.S. EPA registration of PAA products through the Office of Pesticide Programs (OPP). The program regulates the use of all pesticides (including disinfectants ranging from household cleaning products to chemicals used in wastewater treatment) in the United States, and it establishes maximum levels for pesticide residues through the Federal Insecticide, Fungicide, and Rodenticide Act (FIFRA) (7 U.S.C. §136 et seq., 1996).

The Federal Insecticide, Fungicide, and Rodenticide Act provides for federal regulation of pesticide distribution, sale, and use. All pesticides distributed or sold in the United States must be registered (licensed) by U.S. EPA. Before U.S. EPA may register a pesticide under FIFRA, the applicant must show, among other things, that using the pesticide according to specifications "will not generally cause unreasonable adverse effects on the environment". The Federal Insecticide, Fungicide, and Rodenticide Act defines the term *unreasonable adverse effects on the environment* to mean "any unreasonable risk to man or the environment, taking into account the economic, social, and environmental costs and benefits of the use of any pesticide".

In considering the development of registration labels that include allowable residuals for discharge, OPP assesses potential human health and environmental effects associated with use of the product. A company that applies for a registration label for any pesticide must provide data from studies that comply with testing guidelines for development of risk assessments. The risk assessments are used to evaluate the potential for harm to humans, wildlife, fish, and plants, including endangered species and non-target organisms and contamination of surface water or groundwater from leaching, runoff, and spray drift.

Pesticide registration and environmental toxicity as related to FIFRA are not typical considerations in the wastewater treatment industry, which defers to the toxicity criteria and standards established by U.S. EPA and states pursuant to CWA. Wastewater discharge permitting authorities and discharge facilities typically do not refer to chemical treatments such as chlorine or PAA as "pesticides", despite their regulation as such. Additionally, the technologies that enable detection of PAA in water, which provide the ability to understand the presence of PAA in discharges and corresponding risks to aquatic life, are still evolving, and U.S. EPA has yet to publish an approved method for PAA measurement and monitoring in 40 C.F.R. 136, as described in Chapter 4. These realities have resulted in states taking a variety of approaches in setting PAA residual limits, or even permitting use of PAA at all. Table 8.1 provides a summary of known state approaches to setting PAA residual limits in various states.

2.2.1 Use of a State-Determined Water Quality Criteria

In many states, the ambient criteria for a PAA residual in discharges is assumed to be 1.0 mg/L PAA for the receiving stream, with some variation depending on mixing zone policies. The 1.0-mg/L PAA criterion is used as

TABLE 8.1 Summary of several known state approaches to regulating PAA residual limits as of 2018.

State	Allowable PAA residual
Alabama	1 mg/L
Arkansas	Up to 2 mg/L, depending on receiving waterbody
California	Allowing trials with quenching; anticipate U.S. EPA label requirements
Colorado	Allowing trials with 0.4 mg/L at DF = 0; up to 0.7 mg/L above DF[a] > 9
Florida	1 mg/L
Georgia	1 mg/L
Illinois	1 mg/L
Iowa	0.7 mg/L as daily maximum, and 0.4 as monthly average; 2.0 mg/L for CSO[b] applications
Kentucky	Per engineer recommendation, generally 1 mg/L
Maryland	Allowing testing at 0.56 mg/L as a monthly average with a single grab maximum of 0.93 mg/L
Missouri	1 mg/L as a daily maximum; 0.7 mg/L as a monthly average
New Jersey	1 mg/L
Ohio	0.33 mg/L for Steubenville, Ohio
Oklahoma	Guidance published August 2017; allowable residual of 1 mg/L
Oregon	1 mg/L, requiring monitoring of enteroviruses
Tennessee	0.4 for low dilution factors (< 12); site specific for high dilution factors (> 12), up to 2 mg/L
Texas	Anticipate U.S. EPA label, one facility in Texas is using PAA, but did not reopen their permit and quenching is provided for this project

[a]DF = dilution factor
[b]CSO = combined sewer overflow

a base case and may be modified based on data that has been submitted in support of a specific application.

For example, the Metropolitan Wastewater Reclamation District in Denver, Colorado, initiated a full-scale pilot project in January 2018 that was based on target PAA effluent residual concentrations of 0.7 mg PAA/L as a daily maximum and 0.4 mg PAA/L as a calendar monthly average. The basis for the targets was an October 2004 Block Environmental Services study titled, *Aquatic Toxicity Testing Results for PERASAN™ (an Equilibrium Mixture of 15% Peracetic Acid [PAA] and 22% Hydrogen Peroxide) EPA Product #63838-2 Acute and Chronic Freshwater Fish, Invertebrate, and Algal Bioassays*. The Block Environmental Services (2004) study is currently the most comprehensive study available that evaluates the aquatic toxicity of PAA and hydrogen peroxide in a flow-through scenario. The study provides information regarding aquatic toxicity and ecological data for an equilibrium mixture of PAA and hydrogen peroxide in an experimental setup that considered constant exposure to PAA for disinfection on *P. promelas* and *C. dubia*.

The type of test used in the Block Environmental Services (2004) study is a flow-through continuous renewal exposure system, which is different from the static exposure system used in 40 C.F.R. 136 method EPA/821/R-02-013. A flow-through continuous exposure system is generally used to generate and maintain predictable toxicant concentrations continuously throughout the duration of the test. The flow-through continuous exposure system has been shown to be helpful for studying a chemical that decomposes rapidly in water. The no-observed effect concentration (NOEC) is the highest concentration of toxicant that has no observable adverse effects on the test organisms. The inhibition concentration is a point estimate of the toxicant concentration that would cause a given percent reduction in a non-lethal biological measurement. The NOEC result of 0.68 mg PAA/L for the more sensitive species (*C. Dubia*), 0.68 mg PAA/L, from the Block Environmental Services (2004) study, was used to inform the daily maximum target PAA effluent residual concentration of 0.7 mg PAA/L. As an additional protective measure, the Metro District selected a second target effluent residual concentration—a monthly calendar average effluent residual concentration of 0.4 mg PAA/L, which is half of the daily maximum (applying significant digits).

As explained in the Colorado Division's Water Pollution Control Program (regulatory permitting agency in Colorado), *Policy No.: WPC Program, Permits – 1, Implementation of the Narrative Standard for Toxicity in Discharge Permits Using Whole Effluent Toxicity (WET) Testing*, dated September 30, 2010 (Colorado Water Quality Control Division, 2010), "WET testing is not the standard for toxicity, but is an approved 40 C.F.R. Part 136 method that the Division intends to use as a tool for implementation of the narrative standard [section 31.11(1), 5 CCR 1002, 31, narrative toxicity standard]" (page 1 of WET Testing Policy). Based on this statement in the Division's WET Testing Policy, it is the Division's stated policy that the mechanism for implementation of the narrative toxicity standard in section 31.11(1) is by including WET testing requirements in a discharge permit. Whole effluent toxicity testing is the Division's established method of implementing the narrative standard and is appropriate to use here to determine toxicity. A discharge that "passes" a WET test is thus protective of section 31.11(1), the narrative toxicity standard.

2.2.2 Use of a Demonstration Approach

In conjunction with the methodology described in Section 2.2.1, some states are also requiring demonstrations of PAA performance before issuance of a final permit for PAA. Typical requirements may include the following:

- Pre-PAA discharge benthic macro-invertebrate survey or index;
- Whole effluent toxicity testing to determine toxic levels in the receiving water in accordance with state protocols;
- Bench-, pilot-, and/or full-scale demonstration testing to determine the efficacy of the disinfection within the PAA concentration restriction imposed by the WET test results in the receiving stream using a specific PAA formulation (some states will allow WET test results from other similar facilities if the same PAA formulation has been successfully tested);
- Test reports and an NPDES permit modification/change request; and
- Potentially, a post-PAA implementation benthic macro-invertebrate survey or index.

2.2.3 Additional Requirements for Hydrogen Peroxide

Because PAA is an equilibrium mixture that also includes hydrogen peroxide, as described in Chapter 2, there are some states that have additional provisions for addressing hydrogen peroxide. Thus, beyond the above scenarios, some states also include the published WQBEL for hydrogen peroxide (H_2O_2) as part of their criteria. This is complicated, however, because many of the toxicity studies that have been conducted to develop data for registration of PAA products have been conducted with the commercially available solutions, which contain various ratios of hydrogen peroxide. In these cases, the toxicity testing already accounts for the hydrogen peroxide and an extra regulatory requirement for hydrogen peroxide confounds the interpretation of the data used for U.S. EPA registration of some these products.

2.2.4 Information Required for Discharge Permitting in the United States

Table 8.2 provides the information and data that may be required in the process of requesting approval of an NPDES permit or test program from a state regulatory agency. This information is generally needed for determining

TABLE 8.2 Information checklist.

	Required information	Purpose
	Engineering considerations for alternatives analysis	
1.a.	Receiving stream characteristics: 7Q10[a], stream classification, recent stream assessments (up- and down-stream of the WRRF discharge).	To calculate dilution factor, if applicable, and assess availability for receipt of additional toxic compounds. If pre- and post-PAA application macro-invertebrate surveys are required.
1.b.	If facility is existing, NPDES permit limits and record of discharge performance.	To calculate impact of PAA on carbonaceous biochemical oxygen demand based on PAA dose for disinfection.
1.c.	Existing volume in the contact tank, and discharge flow patterns.	To calculate contact time for disinfection (generally, if contact basins suffice for chloramine or free chlorine, contact time will be adequate for PAA disinfection).
1.d.	PAA formulation and recommended feed rate; specifications on feed system.	To calculate proportional feed rates over range of wastewater flows; size of pumps and day storage as well as bulk storage; ensure storage is safe and sufficient to meet feed rate and transportation frequencies; compatibility of feed system to PAA formulation.
1.e.	Cost of PAA and alternative disinfectants.	To calculate cost comparisons between alternative systems.
1.f.	Operation and maintenance requirements for PAA and alternative systems.	To calculate cost comparisons between alternative systems.
1.g.	Capital costs of alternative installations.	To calculate cost comparisons between alternative systems.
	Consideration for protection of water quality	
2.a.	Conduct macro-invertebrate survey, if appropriate. Macroinvertebrate sampling and scoring must follow state-specific protocols. Include information on sample locations and time of year the survey was conducted.	Determination of long-term effect on receiving water-body.
2.b.	Review state protocols for WET testing. Comply with protocols.	To determine allowable concentration of PAA in stream.

(*continued*)

TABLE 8.2 Information checklist (Continued).

	Required information	Purpose
	Consideration for protection of water quality	
2.c.	Apply dilution factor to calculate end of pipe/end of contact chamber concentration of PAA.	To prevent toxicity in stream.
	Alternatives analysis and reporting to state regulatory agency	
3.a.	Prepare plans, specifications, test procedures and data requirements for pilot test.	For NPDES permit modification or pilot approval from state regulatory agency.
3.b.	Collect data on applicable indicator organism to meet disinfection compliance using PAA.	To provide argument for/against adequacy of PAA as a disinfection agent within non-toxic discharge requirements.
3.c.	Apply economic factors from Step 1 to calculate cost-effectiveness of PAA compared alternatives.	To provide argument for permanent NPDES permit modification to PAA in report to state regulatory agency.
3.d.	Conduct macro-invertebrate survey, if required, at same sampling locations and time of the year/flow conditions as pre-pilot survey.	Compare pre- and post-pilot survey results to provide support for/against permanent NPDES permit modification to PAA in report to state regulatory agency.

[a]7Q10 refers to the 7-day low flow in a 10-year period of record

the cost-effectiveness and feasibility of the implementation of a PAA system. The information checklist provided assumes that the stream is "wadable" (or has a small available dilution factor). Larger, "non-wadable" (i.e., larger dilution factor) streams may not require the biological data collection and analyses because there is typically a larger dilution factor available, which reduces the effective residual concentration in the receiving waterbody.

2.3 Peracetic Acid Disinfection Permitting in Canada

Similar to the United States, Canada registers pesticides through the Health Canada Pest Management Regulatory Act (PMRA), under the authority of the Pest Control Products Act and related regulations. The process requires an evaluation of available scientific information that, under the approved conditions of use, demonstrates that the products have value and do not present an unacceptable risk to human health or the environment. To reach decisions related to registration of products like PAA, PMRA applies rigorous risk-assessment methods and policies. These methods consider the unique characteristics of sensitive subpopulations in humans as well as organisms in the environment. These methods also consider the nature of toxic responses and the uncertainties when predicting the impact of pesticides (healthcanada.gc.ca/prma). Several PAA products have been registered under PMRA and have been implemented at several facilities across Canada.

Like the United States, Canada has CCME (2007) protocols for the derivation of Canadian water quality guidelines (WQG), including methodologies for developing acute benchmarks for chemicals for ambient water quality. This approach promotes the use of probabilistic Species Sensitivity Distributions (SSDs) for derivation of guidelines, where enough ecotoxicity data are available. The SSD approach takes all acceptable ecotoxicity data into account in the derivation of the guideline value. However, many provinces do not have an approved process to develop acute water quality criteria for use in effluent limits. Generally, effluent limits are derived from accepted chronic-based water quality criteria to be met at the edge of the mixing zone and back-calculated to an effluent value for use in an Environmental Compliance Approval. In addition, effluent discharges are expected to be nontoxic based on standard acute toxicity tests.

In British Columbia, PAA was introduced to the Canadian municipal wastewater treatment market at the Northwest Langley WRRF, where it was trialed as early as 2012. The approach to permitting at this facility included coordination with the provincial and federal government. Representatives from the British Columbia Ministry of Environment visited the Northwest Langley WRRF in February 2013 to view the disinfection system and discuss the testing program and its objectives. The Ministry of Environment's coordination with the Department of Fisheries and Oceans (DFO) and Environment Canada yielded feedback indicating that as long as materials being used in the effluent for disinfection were nontoxic, neither DFO nor Environment Canada were mandated to, or had resources for, evaluating municipal effluent disinfection issues (Golder Associates, 2013).

The Ontario Ministry of the Environment, for example, has had concerns with the traditional approach for PAA because it degrades rapidly in the environment, and the typical chronic toxicity approach may not be valid. Thus, understanding that there may be difficulties in carrying out acute toxicity tests, the Ontario Ministry of the Environment has evaluated alternative numerical benchmarks that could be considered as non-acutely toxic. The approach used by the Ontario Ministry of the Environment leverages a dataset, which includes toxicity information for marine species. This is consistent with the approaches of both the Canadian government (Government of Canada, n.d.) and the European Union, which derive freshwater quality criteria by combining toxicity data for saltwater and freshwater species to fill in minimum dataset requirements. This approach is specific to organics, where it has been shown that there is no known difference in toxicity between saltwater and freshwater species, so data may be combined. Inclusion of marine data allows for derivation of a water quality criteria (WQC) for PAA using the SSD approach. Ontario has derived a draft WQC (200 μg/L) value using the probabilistic SSD approach (personal communication Vince Pileggi, Sr. Wastewater Engineering Advisor, Water Standards Section, Standards Development Branch, ESSD, MOECC, January 2018).

3.0 EDUCATION, TRAINING, AND KNOWLEDGE SHARING

Until uniform standards are adopted, it will be important to collaborate with other regulatory agencies experienced in activities related to testing and approving PAA products. This knowledge sharing can support development of site-specific test procedures, which may be developed over time. At the same time, it is anticipated that referenced professional, technical, and industry journal PAA articles will continue to increase in publication of related materials. It is important to keep abreast of developments and such research will allow utilities to save time and money when it comes to considering the adaption or conversion of other disinfectants to PAA.

At this point, more than half of U.S. states have been involved in evaluating requests from their activities for PAA approval or for at least a pilot facility to determine the dosage of PAA required to meet disinfection. Similar levels of interest are apparent in Canada. However, protocols for testing to determine the effectiveness of PAA as a disinfectant or for protection of the receiving stream is changing as the national experience increases. Therefore, engaging the regulatory agency as early as possible is important.

It is critical for wastewater discharge permit applicants to request protocols and procedures for testing or implementing PAA, particularly if PAA has not previously been permitted under the specific regulatory authority. Because there can be regulatory uncertainty, it is important to gather as much information as possible on a specific formulation of PAA, particularly if it has not already been permitted. Even if PAA has been approved for application under a particular regulatory authority, this approval process will still most likely require site-specific testing, although data from similar sites may aid in refining the initial permitting process, and this usually saves time. Most importantly, if biological assessments are required, an early start reduces delays in waiting for acceptable circumstances for these assessments and can save a year or more in the process. Engaging the regulatory authority early can streamline the permitting process by helping the agency obtain the required data to issue a permit.

4.0 REFERENCES

Bell, K. Y., & Wylie, V. (2016). The age of peracetic acid—A solution to increasingly challenging regulations. *WaterOnline*, January, 16–18.

Block Environmental Services. (2004). Aquatic Toxicity Testing Results for PERASAN™ (an Equilibrium Mixture of 15% Peracetic Acid [PAA] and 22% Hydrogen Peroxide) EPA Product #63838-2 Acute and Chronic Freshwater Fish, Invertebrate, and Algal Bioassays.

Colorado Water Quality Control Division. (2010). *Implementation of the narrative standard for toxicity in discharge permits using whole effluent toxicity (WET) testing*. Retrieved from https://www.colorado.gov/pacific/sites/default/files/WQ_WET-Testing-Policy.pdf

Deeb, R. A., Dzombak, D., Theis, T., Ellgas, W., & Kavanaugh, M. (2003). The cyanide challenge. *Water Environment & Technology, 15*(2), 35–38.

Golder Associates. (2013). Literature review and results of pilot trials at Langley WWTP—Performance of peracetic acid as an alternative disinfectant. Submitted to Metro Vancouver, WWTP Operations, Delta, BC.

Government of Canada. (n.d.). Chemicals Management Plan – Federal Environmental Quality Guidelines. https://www.canada.ca/en/health-canada/services/chemical-substances/factsheets/federal-environmental-quality-guidelines.html

Pandit, A., Young, C., Pang, M., Khoury, J., Carr, S., Fischer, D., & Stahl, J. (2006). Factors affecting cyanide generation in chlorinated wastewater effluent matrix. *Proceedings of the Water Environment Federation 2006*(7), 4957–4970.

Stephen, C. E., Mount, D., Hansen, D., Gentile, J. R., Chapman, G. A., & Brungs, W. A. (1985). Guidelines for deriving numerical national water quality criteria for the protection of aquatic organisms and their uses. U.S. Environmental Protection Agency.

Švec, P., & Devriese, L. A. (2009). Genus *enterococcus*. In D. E. Vos P, G. M. Garrity, D. Jones, N. R. Krieg, W. Ludwig, F. A. Rainey, K.-H. Schleifer, & W. B. Whitman (Eds.). *Bergey's manual of systematic bacteriology* (2nd ed., Vol. 3, pp. 594–607). New York, NY: Springer.

U.S. Environmental Protection Agency. (1991a). *Methods for measuring the acute toxicity of effluents to aquafic organisms* (4th ed., EPA-600/4-90-027). Cincinnati, OH: U.S. Environmental Protection Agency, Office of Research and Development.

U.S. Environmental Protection Agency. (1991b). *Short-term methods for estimating the chronic toxicity of effluents and receiving waters to freshwater organisms* (3rd ed., EPA-600/4-91/002). Cincinnati, OH: Environmental Protection Agency U.S. Environmental Protection Agency, Office of Research and Development.

U.S. Environmental Protection Agency. (2008). Memorandum from Ephriam King, Director EPA Office of Science and Technology to William Spratlin, Director Water, Wetlands and Pesticides on Initial Zones of Dilution for Bacteria in Rivers and Streams Designated for Primary Contact Recreation.

U.S. Environmental Protection Agency. (2015). Ambient WQ criteria. Retrieved from: https://www.epa.gov/wqc/2015-epa-updated-ambient-water-quality-criteria-protection-human-health

Wang, W. (1990). Toxicity assessment of pretreated industrial wastewaters using higher plants. *Research Journal of the Water Pollution Control Federation, 62*. 853–860.

Weng, S.-C., Dunkin, N., Schwab, K. J., McQuarrie, J., Bell, K., & Jacangelo, J. G. (2018). Infectivity reduction efficacy of UV irradiation and peracetic acid-UV combined treatment on MS2 bacteriophage and murine norovirus in secondary wastewater effluent. *Journal of Environmental Management*, 221. 1–9.

Zheng, A., Dzombak, D. A., & Luthy, R. G. (2004). Effects of thiocyanate on the formation of free cyanide during chlorination and ultraviolet disinfection of publicly owned treatment works secondary effluent. *Water Environment Research*, 76(3), 205–212.

9

Design and Process Control Considerations

Sarah Stewart, Leonard Levine, Brian Hilts, Samuel Lowen, & Alberto Garibi

1.0 INTRODUCTION

The design and construction of a peracetic acid (PAA) system can be provided by any of the normal project delivery methods, such as

- design/bid/build,
- construction-manager-at-risk,
- design/build, and
- other methods.

Independent of the method chosen, it is important that several key decisions be made very early in the process and that owner input from operations, maintenance, management, and engineering staff be given proper weight in the decision-making process. Making and adhering to these decisions will contribute greatly to a cost- and schedule-efficient project delivery.

It is highly recommended that the PAA provider be responsible for construction oversight and sign-off of a PAA system, especially if they will be responsible for the overall design and implementation of the system and ongoing maintenance throughout the length of the lease agreement. It is also prudent that the overall design engineer and general contractor (depending on the contract type) be involved in construction oversight. Finally, the owner and owner's representatives, including the quality assurance/quality control subcontractor, must be involved in the oversight to make sure the owner receives the system specified and shown in the contract documents.

1.1 Renovating Existing Infrastructure

Using renovated existing structures for installation of a new PAA system will typically reduce capital costs and project schedule. It will most certainly reduce the impact on available site area. The challenges of reusing existing structures for a new PAA system come in the implementation of the change from the existing disinfection process to a PAA system.

When considering this option, it must be determined what part, if any, of the existing infrastructure can be used without jeopardizing the current operations, maintenance, and compliance. A key part of the decision-making process should be the development of a transition plan that considers such factors as temporary operating conditions and maintaining permit compliance. In some cases, the plan will rely on certain weather conditions to be viable.

Other considerations include the condition of the current infrastructure, its projected lifespan, and what will need to be modified to use PAA instead of the existing disinfection process. Depending on available capital, a decision may be made to start with existing infrastructure and install new structures as part of the facility's long-term capital planning. If this approach is selected, it is recommended that the PAA feed and storage equipment be located with the new location in mind.

1.2 All New Facility

Designing a new PAA system eliminates the need to worry about the condition and lifespan of the current infrastructure. If done correctly, there is less concern and planning for maintaining current operations and maintenance and permit compliance. This option will likely come with a higher capital cost, take up a larger footprint, and may result in a longer project schedule. Technical considerations for this approach include installing and starting up the new PAA system while maintaining operability, maintainability, and permit compliance. Also, installers must ensure that the new flow configuration will not reduce the hydraulic capacity of the facility.

2.0 SYSTEM COMPONENTS

2.1 Contact Tank

For PAA to achieve the desired disinfection efficacy, a site-specific contact time must be provided. Contact time can be achieved through a contact tank or through effluent conduits such as open channels or piping. Most applications require a contact tank; therefore, this section will look at aspects and components of, and considerations for, a properly designed contact tank. A properly designed contact tank for PAA disinfection will be similar to one designed for chlorine disinfection.

2.1.1 Mixing

A critical component of achieving the desired disinfection efficacy at a water resource recovery facility (WRRF) is to maximize the use of the contact tank volume; therefore, obtaining good mixing of PAA into the effluent is essential. The following types of mixing are available: static, mechanical, and hydraulic.

It is important to note that carrier water should not be used to feed PAA because the additional water will push the equilibrium reaction and convert some PAA back into acetic acid and hydrogen peroxide. Therefore, it is recommended to feed PAA at the commercially available strength (e.g., 15%).

2.1.1.1 Static

Static mixing is accomplished by adding PAA upstream of a fixed element, such as a static mixer or baffle walls, to generate a homogenous mixture. Pressure drop across the fixed elements is required to generate the necessary mixing energy. Figure 9.1 presents an example baffle for mixing, and the following summarizes advantages and disadvantages of this type of mixing system:

Advantages	Disadvantages
• Low maintenance	• Generates pressure drop or headloss along the facility hydraulic profile, which may reduce the hydraulic capacity of the facility
• Low pump energy required	• Reduced efficiency at low flows
	• Floatables, if present, are captured and must be removed/managed

FIGURE 9.1 Example of baffle wall for mixing (courtesy of the City of Memphis, Tennessee).

2.1.1.2 Mechanical Mixing

Mechanical mixing uses mechanical equipment to impart energy into the flow stream to achieve a well-mixed effluent. Two main types of mechanical mixing are available: induction mixing and mechanical mixers. In the case of induction mixing, a submersible motor and induction unit is located in an open channel or tank, along with a guide rail and hoisting mechanism, as seen in Figure 9.2. Peracetic acid enters the unit, and mixing energy is imparted as the PAA is mixed with the effluent. For mechanical mixers, PAA is added to the effluent immediately upstream of the mixer, which, in turn, adds energy to the effluent to generate a homogenous mixture. The following are the advantages and disadvantages of this mixing approach:

Advantages	Disadvantages
• Properly sized systems generate completely mixed effluents	• Requires maintenance (shelf spare is recommended to allow continuous operation during repairs)
• Less headloss compared to static or hydraulic mixing options	• Capital cost can be high relative to other alternatives
• Provides good mixing over a wide range of flowrates	• Higher energy costs compared to other mixing alternatives

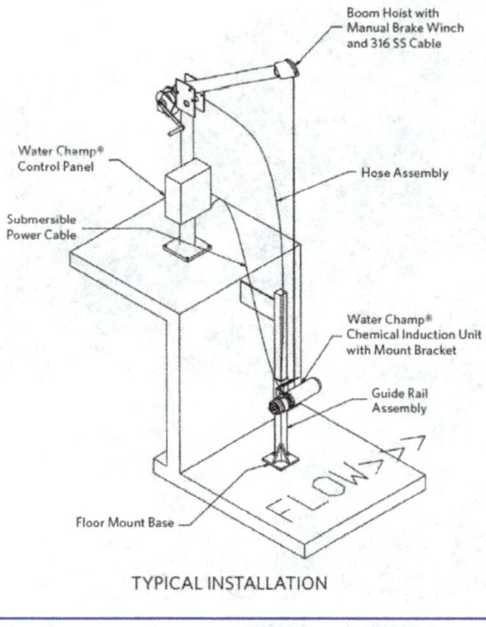

FIGURE 9.2 Typical induction mixer installation (courtesy of Evoqua).

2.1.1.3 Hydraulic Mixing

Hydraulic mixing uses a hydraulic element in the facility hydraulic profile to generate a well-mixed effluent. Ideally, this is accomplished with an existing hydraulic element such as a weir, flume, or hydraulic fall in the facility, with PAA being added immediately upstream of the element. Figure 9.3 shows an example of hydraulic mixing. The advantages and disadvantages of hydraulic mixing are as follows:

Advantages	Disadvantages
• No additional energy is required	• May not be possible if insufficient headloss is available in the existing hydraulic profile
• Low maintenance	• May be less efficient at low flows
• Generates completely mixed effluent	

FIGURE 9.3 Example of hydraulic mixing (Reprinted with permission from City Water and Light of Jonesboro, Arkansas).

2.1.1.4 Computational Fluid Dynamics

Computational fluid dynamics (CFD) is an effective tool that can be used to assess the mixing generated by the selected mixing approach. Software is used to mimic the conditions at a given site, and it will determine how well the effluent is mixed at various flow conditions. Computation fluid dynamics analysis is strongly recommended if static mixing is being used for applying PAA to an effluent. Figure 9.4 illustrates the output of a CFD model, where the color gradient represents the percent deviation of PAA concentration from the target concentration.

2.1.2 Tank Geometry

The geometry of a contact tank has a large impact on its effectiveness. Peracetic acid contact tanks should be designed in a similar fashion to those used for chlorine disinfection. Plug flow conditions generate best results compared to a completely mixed tank. As a result, the contact tank should have a length-to-width ratio of at least 40:1, according to Metcalf & Eddy (2003), while larger length-to-width ratios can also be used. The height-to-width ratio should be 1:1 or less. Another key element to tank geometry is a symmetrical design of the tank to generate an even flow split should there be two halves of the contact tank, which is advised.

2.1.3 Functional Considerations

There are numerous functional aspects of a contact tank design that should be considered.

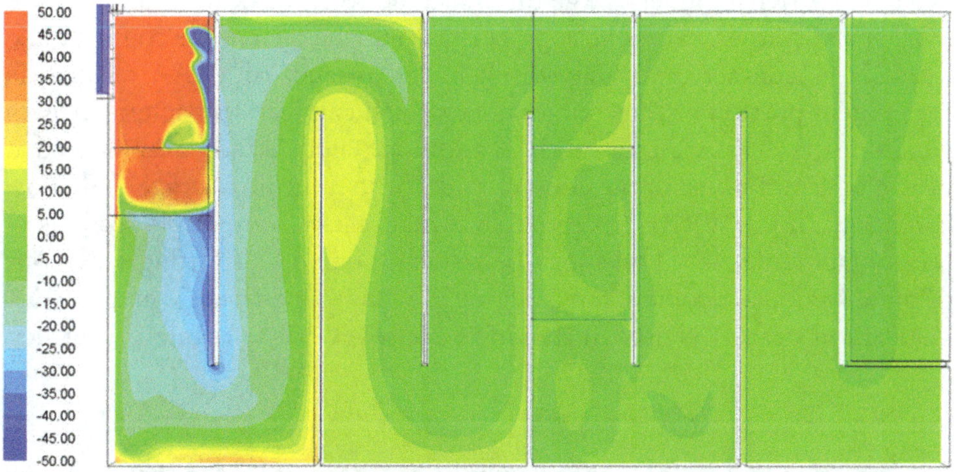

FIGURE 9.4 Example of CFD model output (courtesy of PeroxyChem).

2.1.3.1 Proper Sizing of the Contact Tank

Determining the proper volume or size of the PAA contact tank is a key element to achieving the desired disinfection performance of a PAA system and should be performed in concert with the selection of the PAA dose. In many cases, existing infrastructure (tanks and/or pipes) are available to be used as part or all of the necessary contact time. The volumetric sizing of a contact tank for PAA disinfection is often similar to that for chlorine disinfection. It is highly recommended that both bench testing and pilot testing be conducted to verify the design for contact time because site-specific conditions may warrant higher or lower contact times. The contact tank may potentially be sized smaller, depending on the bench- and pilot-scale test results. Minimizing contact times will result in lower capital costs and a smaller footprint if those factors are important to the owner.

The lower contact time may also result in higher operating costs, which would be determined from the batch and pilot testing. The owner would then weigh the capital cost and smaller footprint versus a potentially lower operating cost. Tanks can be sized larger, but care must be taken to make certain that a PAA residual is maintained throughout the contact tank, otherwise bacteria growth may occur. It is worth noting that some regulatory agencies require minimum contact times.

2.1.3.2 Peracetic Acid Application Points

Peracetic acid application locations should be considered during the design of a PAA disinfection system. Typically, one application point is provided to meet the design criteria for contact time at peak hour flows. However, if a WRRF experiences a very large difference between peak hour and current average daily flows, excessive contact times may be seen in the contact tank. These large contact times may result in the absence of a PAA residual in the effluent portion of the contact tank, which can lead to bacteria growth conditions. As a result, a second PAA addition point should be considered at or near the mid-point of the contact tank for use during average daily flow conditions. If the WRRF has multiple contact tanks or multiple sections of the contact tanks, low flows may be accommodated by shutting down one or more sections or contact tanks during low flow conditions. Whatever PAA application strategy is used, it should be approved by the regulatory agency.

2.1.3.3 Isolation and Drainage

For operation and maintenance flexibility, it is recommended that a contact tank system have two or more sections. Each section of the contact tank should include a means of isolation at the influent and effluent ends and a means of draining the contact tank. Isolation is typically achieved using a valve, gate, or stop logs/plate at the influent end of the contact tank. The effluent end of the contact tank typically has a weir, which provides isolation for the tank. Tank drainage should be provided via mud valves or other means. Tank drains should be sent to the in-plant drain system. The tank floor should be sloped to the tank drains to facilitate cleaning of the tank.

2.1.3.4 Access and Maintenance

The contact tank should include well-thought-out access points to allow for the operation and maintenance of the disinfection system and contact tank. Typically, access is provided around the perimeter of the tank and down the center of the tank (above the tank divider wall if the tank has two halves). Additional access should be provided to locations that require maintenance, which may include PAA application/mixing points, sample locations, flow or level elements, PAA residual analyzers, compliance samplers, and locations to wash down the tank. Wash hose stations should be placed at appropriate locations so the tank can be properly cleaned when it is offline.

2.1.3.5 Associated Appurtenances

Several other appurtenances should be considered when designing a contact tank for PAA disinfection. These include the following:

- Pressure-relief valves—High groundwater levels may warrant the use of pressure-relief valves to mitigate contact tank buoyancy issues.
- Guardrail—Safety requirements mandate the use of guardrails and toeplates on the contact tank.
- Floatable removal—Each site should evaluate the need for including floatable removal. If in doubt, it is wise to include floatable control.
- Contact tank covers—Some sites might do well by adding covers to the contact tanks to minimize algae growth.
- Concrete coating—While most sites find this unnecessary, some sites may want to include a concrete coating system at the water line of the contact tank to help prevent algae from attaching to tank walls and to facilitate cleaning.

2.2 Chemical Storage

2.2.1 Code Review and Requirements

The properties of PAA, as described in Chapter 2, allow it to be stored either inside or outside; there are applicable codes for storing PAA depending on whether the project design includes chemical storage in a building or outside. A complete building code key determinations review should be performed during the 10% design phase and should be reviewed with the authority having jurisdiction for approval. In general, however, PAA storage and feed buildings will be classified as a Group H-3 and H-4 occupancy if they exceed the indoor storage limits allowed by the following codes:

- 2012 International Building Code (IBC),
- 2012 International Fire Code (IFC), and
- 2012 International Mechanical Code (IMC).

2.2.1.1 Indoor Storage

Peracetic acid is classified in accordance with IFC as possessing the following physical properties.

- Combustible Liquid (Class III-A)—H-3 occupancy if over 1250 L (330 gal)
- Organic Peroxide (Class IV)
- Oxidizer Solid/Liquid (Class II)—H-3 occupancy if over 95 L (25 gal)

In addition, PAA is also classified in accordance with IFC as possessing the following health properties:

- Corrosive—Requires H-4 occupancy if chemical volume is over 1890 L (500 gal)
- Toxic—Requires H-4 occupancy if chemical volume is over 190 L (50 gal)

The maximum allowable quantities per control area (no sprinklers installed) listed are based on Table 3.07.1(2) and Section 307.5, High-hazard Group H-3. The maximum allowable quantity per control area can increase 100% if the building has sprinklers.

Within a closed system, if the amount of PAA exceeds the amounts listed above, H-3 and H-4 occupancy classification will be triggered.

Group H-4 requirements are as follows:

- Install and maintain automatic fire detection system.
- Install and maintain automatic fire sprinkler system.
- Ensure presence of mechanical ventilation with ducts extending directly to the exterior of the building without entering other spaces.

Ventilation system requires a manual shutoff control outside of the room (IBC Section 414.3, Ventilation).

- Locate exhaust outlets and ducts discharging to the outdoors
  - 3 m (10 ft) from the property lines,
  - 1 m (3 ft) from exterior walls and roofs,
  - 3 m (10 ft) from operable openings into buildings, and
  - 3 m (10 ft) from adjoining grade.
- For other product-conveying outlets:
  - Monitor control equipment (IBC Section 415.5.2)

- Ensure presence of emergency or standby power for the mechanical ventilation system (IBC Section 415.5.3).
- Ensure presence of secondary spill control and containment system (IBC Section 415.5.4).
- Ensure presence of emergency alarms for the detection and notification of an emergency condition. Supervision and monitoring of these alarms is also required (IBC Section 414.7).
- Ensure floors are of liquid-tight, noncombustible construction (IBC Sections 415.7.5 and 415.9.3).
- Maintain a fire separation distance from lot lines and structures of between 0 and 15 m (0 and 50 ft), depending on the type of construction and fire-rating of exterior walls (IBC Section 415.5.3, Table 415.5.2 and Table 602).

The additional elements to satisfy Group H-3 requirements are as follows:

- Locate mechanical ventilation exhaust outlets and ducts discharging to the outdoors (IMC Section 501.3.1)
  - 9.1 m (30 ft) from property lines;
  - 3 m (10 ft) from operable openings in buildings;
  - 1.8 m (6 ft) from exterior walls and roofs;
  - 9.1 m (30 ft) from combustible walls and operable openings into buildings, which are in the direction of the exhaust discharge; and
  - 3 m (10 ft) above adjoining grade for flammable vapors, fumes, or dusts.
- Ensure that not less than 25% of the perimeter wall of the occupancy is exterior wall (IBC Section 415.5).
  - Exception: Liquid storage rooms with area less than 93 m^2 (1000 ft^2) do not need to be located on the outer perimeter, provided they meet IFC and National Fire Protection Association (NFPA) 30 requirements.

- Ensure that floors are of liquid-tight, noncombustible construction (IBC Section 415.9.3).
- Ensure that storage tanks are noncombustible and protected from physical damage. A combination of fire barriers and/or horizontal assemblies located around the storage tanks shall be permitted as the method of protection from physical damage (IBC Section 415.8.2.2).
- Install double-wall storage tanks or a liquid-tight leakage containment area (IBC Section 415.8.2.4).
- Ensure presence of a supervised automatic leakage alarm (IBC Section 415.8.2.5).
- Ensure that storage tank vents terminate to the outdoor air. However, it is recommended to check with your jurisdiction to see if a scrubber is required for the tank vent (IBC Section 415.8.2.6).

2.2.1.2 Outdoor Storage

When PAA is stored outside, the following requirements may apply:

- If weather protection is provided for sheltering outdoor hazardous material storage:
  - Walls shall not obstruct more than one side of the structure (IBC Section 414.6.1.1).
    - Exception: Walls shall be permitted to obstruct portions of multiple sides of the structure, provided that the obstructed wall area is not greater than 25% of the overall possible wall area of the structure.
  - The overhead structure shall be of approved noncombustible construction with a maximum area of 140 m^2 (1500 ft^2).
    - Exception: The maximum area is permitted to be increased by IBC Section 506.
- Outdoor storage or use of corrosives or toxic liquids must be more than 6 m (20 ft) from lot lines, public streets, public alleys, public ways, exit discharges, or exterior wall openings. A 2-hour fire barrier without openings or penetrations extending not less than 0.76 m (30 in.) above and to the sides of the storage is allowed in lieu of such distance. The wall must be either an independent structure or the exterior wall of the building adjacent to the storage area (IFC Sections 5404.2.2 and 6003.2.3.1).
- The area surrounding an outdoor storage area or tank shall be kept clear of combustible materials and vegetation for a minimum distance of 7.6 m (25 ft) (IFC Section 5004.11).
- Secondary spill control and containment system (IFC Table 5004.2.2).

2.2.2 Storage Volume

To determine appropriate PAA storage volumes, the owner should consider minimum inventory, regulatory, and, if applicable, bulk storage requirements. Minimum inventory volume should be based on the amount of PAA necessary to meet disinfection performance at average flow and average dose for at least 14 days, plus the expected delivery time (in days) from the PAA provider. The owner may consider adding additional storage volume to account for possible delays in shipping resulting from unforeseen delivery problems that might be caused by distance from the manufacturing facility to the WRRF and weather conditions.

While the number of PAA suppliers is increasing, there are still not many suppliers compared to other commodity chemicals, so there is some level of risk associated with potential delivery delays. The owner should consider having the PAA provider store an additional amount of PAA near the owner's site to reduce the impact of delivery issues, especially given that PAA has a long shelf life. For example, a facility with an annual average daily flow of 76 ML/d (20 mgd) and an average PAA dose of 2 mg/L as PAA, that procures PAA from a supplier with a 5-day lead time should consider setting an inventory of at least 16 600 L (4380 gal). This is determined by a series of calculations, as follows:

Daily average usage (kg/d) = (76 ML/d)(2 mg/L)/(0.15 concentration of PAA) = 1013 kg/d (2229 lb/day)

Daily average usage (L/d) = (1013 kg/d)/(1.16 kg/L) = 873 L/day (230 gal/day)

Minimum inventory (L) = (873 L/d)(14 + 5 days) = 16 600 L (4380 gal)

Inventory calculations should also consider the total amount of chemical necessary to meet disinfection performance during a peak event, which is based on the daily peak flow, the dose required during peak flow, and the number of consecutive days that the peak event typically lasts. For example, a facility that experiences a peak flow of 230 ML/d (60 mgd) for 3 consecutive days and requires a dose of 3 ppm (mg/L) should maintain an inventory of at least 11 900 L (3140 gal), as calculated by the following:

Daily peak usage (kg/d) = (230 ML/d)(3 mg/L)/0.15 = 4600 kg/d (10 100 lb/day)

Peak event usage (kg) = (4600 kg/d)(3 days) = 13 800 kg (30 360 lb)

Peak event usage (L) = (13 800 kg)/(1.16 kg/L) = 11 900 L (3140 gal)

In this example, because the minimum inventory required for average conditions (16 600 L [4370 gal]) is also sufficient to cover a peak event (11 900 L [3140 gal]), the minimum inventory should be the basis to determine total storage volume. However, if the peak event in the example lasted for 5 days, the chemical amounts consumed during a peak event would be the best criteria to determine the total storage volume.

Another criterion to determine the total storage volume requirement is whether PAA will be supplied in 1250-L (330-gal) totes or in 15 100-L (4000-gal) bulk deliveries. Storage of PAA in totes does not impact determination of total storage volume, although total space requirements are important and discussed in the next section. Storage of PAA in bulk requires installation of storage tanks appropriately sized to safely receive a truckload delivery. As a minimum, storage tanks should be sized with a working volume of 125% of the delivery volume, plus 10% heel and 10% head space. A working volume of 18 900 L (15 100 L × 1.25), a heel volume of 1900 L, and a head space volume of 1900 L, add up to a total minimum tank volume of 22 700 L (6000 gal). In the case of the facility referenced in the aforementioned examples, while the minimum inventory is only 16 600 L (4380 gal), the total storage volume (volume plus tank heel) would be 20 800 L (5500 gal) stored in a 22 700-L (6000-gal) tank. Figure 9.5 shows an installation with passivated stainless steel storage tanks, and Figure 9.6 shows installations with high-density polyethylene (HDPE) storage tanks.

FIGURE 9.5 Passivated stainless steel storage tanks (courtesy of PeroxyChem).

FIGURE 9.6 High-density polyethylene storage tanks (courtesy of PeroxyChem).

2.2.3 International Building Code Tote Storage Considerations

Peracetic acid can be delivered in plastic IBC totes (1250 L [330-gal]). Tote storage areas should meet the following criteria:

- Ambient temperature is normally acceptable. Peracetic acid freezes at −49 °C (−56 °F) and does not require freeze protection. Peracetic acid may undergo decomposition if its temperature reaches 57 °C (135 °F). For long-term storage (more than 90 days), it is recommended to maintain storage area temperatures below 40 °C (104 °F) at all times to minimize degradation. Nearby heat sources that could raise the temperature of PAA totes should be avoided.

- Direct exposure to sunlight should be avoided to prevent embrittlement of the tote's plastic. Owners should keep track and monitor the expiration date of each tote and avoid manipulating full totes after expiration because the integrity of containment may be compromised. Peracetic acid providers should be contacted to remove the tote from the site safely.
- If stored indoors, the storage area should be sufficiently ventilated to avoid accumulation of PAA vapors in case of a spill. Owners should consult Occupational Safety and Health Administration (OSHA) and local regulations to determine if maximum PAA concentration limits exist and whether PAA vapor detectors are required. Given its strong vinegar-like odor, the presence of PAA vapors in very small concentrations is easily detected by operators.
- Storage areas should have access to water (hose) for spill dilution and wash-down and a drain to a location that can safely receive diluted/decomposed PAA (diluted acetic acid). It is recommended that a spill scenario be investigated to determine any potential effect on equipment and/or a biological system, if applicable. Storage of IBC totes does not require additional containment as long as totes remain closed and sealed. Refer to the "Chemical Feed" section for containment requirements for IBC tote feed systems.
- Storage areas should be clearly delineated and marked as storage for strong oxidants. No wooden pallets shall be used for PAA storage or allowed in the PAA storage area. No reducing agents or materials should be stored near the PAA storage area. Additionally, no combustible materials should be stored near the PAA storage area.
- Tote vent caps should not be blocked. It is not recommended to double-stack PAA totes during storage. Totes should be kept upright at all times.
- Storage area layout should allow for safe maneuvering of forklifts.

Figure 9.7 illustrates a typical IBC tote-based PAA storage system.

2.2.4 Bulk Storage Considerations

Bulk storage systems must be designed to ensure safety during PAA transfer from trucks to tanks and from tanks to the application points. The following criteria should be considered for PAA bulk storage.

FIGURE 9.7 Storage totes (courtesy of PeroxyChem).

2.2.4.1 Tank Materials of Construction

Both HDPE and passivated stainless steel storage tanks can be used for storage of PAA. It is recommended that plastic (HDPE) tanks be supplied with double-wall secondary containment systems, making them ideally suited for temporary installations. Single-wall HDPE tanks are acceptable when the tank is installed inside a properly sized containment area. Linear HDPE is preferred over cross-linked HDPE because premature embrittlement of the latter has been observed in the field. Although plastic tanks are typically the lowest cost option, owners should consider the following:

- Plastic tanks are not rigid. The tank shell cannot be used for piping support. Flexible connections (typically stainless steel braided hoses) are required between tank nozzles and piping; top nozzles may lose vertical alignment, which may interfere with radar or ultrasonic level sensors.

- Plastic tanks are typically available as flat-bottom, domed-top configurations. This tank configuration does not allow for full gravity drain of the tank, which is required during commissioning and may be required during maintenance.
- Plastic tanks have a shorter life, typically between 5 and 7 years.
- Tiedowns are typically required to meet wind speed requirements.

Stainless steel tanks are typically more expensive; however, they have a much longer life (15 years minimum) and can be procured with slanted or conical bottoms that allow for gravity drain. Stainless steel tanks can minimize the need for piping supports and enable the option of locating instruments at operator level, eliminating the need for operators to access the top of the tank. All stainless steel equipment must be passivated before PAA contact.

2.2.4.2 Venting and Pressure-Relief

Bulk storage tanks must be equipped with venting and pressure-relief devices that allow the tank to breathe and prevent contaminants from entering the tank, operator exposure to PAA vapors, and damage caused by over-pressurization in a decomposition event.

Pressure-relief (floating) manways must be provided to prevent damage caused by over-pressurization in a decomposition event. The number, size (diameter), and location of the manways shall be determined and documented based on the appropriate PAA decomposition kinetics. The pressure-relief setpoint must be based on the tank's maximum design pressure.

Conservation vents must be provided to allow the tank to "breathe" during normal operation. Conservation vents have one inlet port that allows air into the tank when a small vacuum (e.g., 2-in. H_2O) is created as PAA is drawn from the tank, and one outlet port that allows PAA vapors to leave the tank whenever a small pressure (e.g., 2-in. H_2O) is created when the interior of an idle tank increases temperature overtime. During all other times the conservation vent is closed, preventing foreign materials from entering the tank and minimizing PAA odors in the storage area. Scrubbers can be mounted on the outlet port of the conservation vent to capture the small amount of PAA vapors released as the tank "breathes". Check with regulatory agencies to determine if conservation vents are required. Figure 9.8 shows an installation of a conservation vent on a storage tank.

Vapor return lines connecting the top of the tank to the top of the truck container(s) must be provided to capture PAA vapors displaced as a truckload is being pumped into the tank. The vacuum created in the truck as the PAA level drops pulls the displaced vapors from the tank and into the empty truck container(s). This approach minimizes PAA vapor releases during tank fills and maintains conservation vents and floating manways closed to prevent entry of foreign objects into the tank. In areas where vapor release to the atmosphere is not of regulatory concern, a tank overflow line can be installed to direct vapors near the PAA injection point in the contact tank.

FIGURE 9.8 Storage tank conservation vent (courtesy of PeroxyChem).

Another option is to install vent lines with either water traps or have the lines terminate just below the water surface in the contact tank. Water traps need to have a means of water being added because evaporation will cause the trap to fail over time. Terminating the vent line below the water level in the contact tank should be done carefully because installing the pipe deep below the water surface will require pressure to be built up in the storage tank before the vapor can be released into the water. Use of duck-billed check valves is not recommended because the vapor from PAA is corrosive to the valve material, and fumes will eventually leak out of the pipe.

The use of goose-neck vents is not recommended because they do not prevent foreign materials from entering the tank, they do allow the release of PAA vapors, and, if clogged, they may collapse the storage tank.

2.2.4.3 Tank Discharge

It is typically recommended to locate the tank discharge nozzle at the bottom of the tank to ensure a flooded suction for the feed pumps. The piping design must ensure that no backflow conditions exist. In cases when the tank is not installed within a containment dike and a double-wall tank is installed, a dip tube can be used to discharge from the top and avoid penetration of the tank shell. In these cases, selection of a feed pump that can provide sufficient suction lift is essential. It is also critical to minimize flanges and potential air-leak points in the line to avoid loss of prime for the feed pump. Figure 9.9 illustrates an isometric of a pump skid used for feeding PAA.

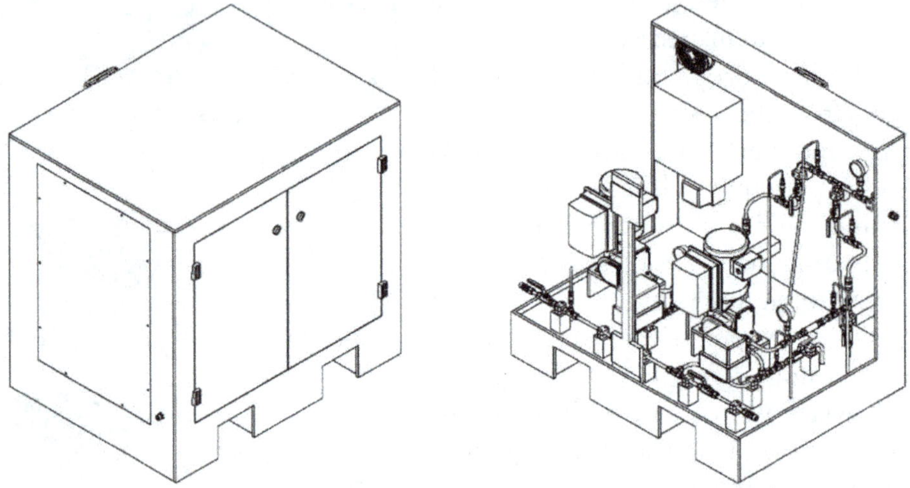

FIGURE 9.9 Pump skid isometric (courtesy of PeroxyChem).

2.2.4.4 Instrumentation

At a minimum, storage tanks should be fitted with the following instrumentation:

- Temperature in PAA storage tanks should be continuously monitored to provide early indication in case of decomposition. When plastic tanks are used, infrared temperature sensors can be installed at the top of the tank. When stainless steel tanks are used, resistance temperature detector (RTD) elements, or thermocouple sensors, installed on the bottom side of the tank can be used. Note that RTDs should penetrate at least 0.3 m (12 in.) inside the tank wall. Temperature shall be displayed continuously, and alarms shall be triggered whenever temperature exceeds 50 °C (122 °F). The PAA provider shall be contacted to provide recommended setpoints for temperature alarms as a function of the PAA formulation in question.
- Level indications should include continuous monitoring, high-level alarm, and a high-high level switch with alarm and safety interlock to shut down the transfer pump filling the tank. When plastic tanks are used, sensors installed at the top of the tank should be used (e.g., ultrasonic or radar elements). When stainless steel tanks are used, additional options include the use of differential pressure elements located at the bottom of the tank. High-high level detection is typically achieved with a different sensing technology. Sight gauges for visual-level indication are not recommended because they represent

an additional point of leak and loss of containment. If the use of a sight gauge is needed, an armored gauge should be required.

- Leak detection (double-wall tanks only)—When double-wall plastic tanks are used, a leak detector should be supplied to alarm operators of loss of containment in the inner shell.

2.2.4.5 Containment

Secondary containment for PAA storage is typically accomplished by installing storage tanks inside secondary containment, such as a concrete curbing or a tank farm with retaining walls. The concrete curbing should be sufficiently high to hold the volume of the largest tank in containment, plus the volume of a 25-year rain event (if located outside) plus a minimum of 150 mm (6 in.) of free board. All concrete surfaces (including the top of the walls) should be coated with a chemical-resistant material compatible with PAA, hydrogen peroxide, and acetic acid. The containment curbing should include a sump pump or gravity sump with an isolation valve to drain the containment area from rainfall events. The operation of the pump or valve should be manual so as to not accidentally pump PAA from an unknown spill. It is important to provide a method to safely remove any spilled PAA from the containment area. Confer with the PAA provider as to how to safely handle a PAA spill condition. At a minimum, spilled PAA should be neutralized before being removed from the containment area.

2.3 Chemical Feed

2.3.1 Pump Type

Most PAA formulations approved for wastewater disinfection contain a high concentration of hydrogen peroxide, which increases off-gassing of oxygen. Therefore, PAA disinfection applications should use a chemical feed pump that provides high levels of reliability, even if heavy off-gassing is taking place. Solenoid and mechanical diaphragm pumps are among the most popular chemical feed pump types used in WRRFs in North America. While several pump vendors offer degassing valves to minimize pump air-binding issues, most wastewater PAA disinfection systems installed in the last several years use peristaltic pumps. Examples of peristaltic pumps that have been successfully used for PAA feed for several years in WRRFs in the United States include Watson Marlow QDOS and Flomotion V2001, which are shown in Figure 9.10.

In addition to pump-type selection, owners should consider materials of construction. Santoprene has been successfully used across the country for peristaltic pump hose/tubing. Useful life of Santoprene tubes depends on rotor speed (revolutions per minute) and frequency of use; in general,

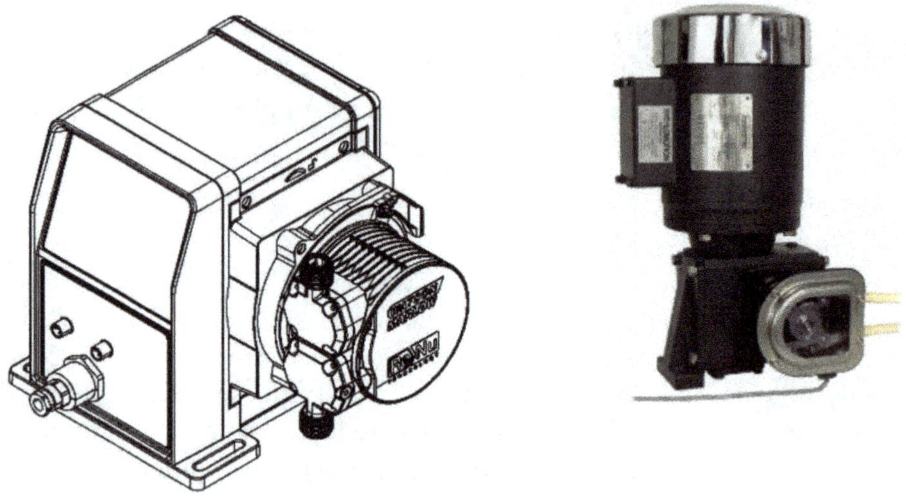

FIGURE 9.10 Example pump types such as Watson Marlow QDOS or Flomotion V2001 (courtesy of PeroxyChem).

30 to 90 days can be expected. Given the importance of reliability in disinfection applications, the use of duty and redundant pumps with automatic switchover capabilities has become a common practice for PAA feed skids for municipal wastewater disinfection.

2.3.2 *Piping and Appurtenances*

The following piping requirements should be considered for all PAA service lines:

When small-bore lines (less than 13-mm [0.5-in.] diameter) are required, Teflon or 316SS tubing with compression fittings may be used. For larger diameter lines, 304 or 316 butt-welded joint piping is recommended. Although the use of flanges in PAA piping systems is necessary for ease of assembly and disassembly, at valve connections, their use should be reduced because it increases the chance of leaks in the piping system. To the extent possible, threaded connections in PAA service should be avoided because of the tendency to leak. The use of flanged stainless steel braided hoses when flexible connections is required.

Pressure-relief valves must be installed in any section of pipe that could entrap PAA. Proper sizing, selection, and documentation of pressure safety valves is required. Relief lines should be routed to containment and must avoid splashing operators standing near the release area. Figure 9.11 shows in installation of a pressure-relief valve.

FIGURE 9.11 Pressure-relief valve installation (courtesy of PeroxyChem).

All ball valves in PAA service lines should be vented by drilling a 3-mm (1/8-in.) hole on the upstream side of the ball. Figure 9.12 shows vented ball valves in the closed (A) and opened (B) positions.

Double-block-and-bleed isolation should be used to prevent critical process backflow or to allow replacement of instruments in critical positions. All stainless steel lines should be fully passivated before contact with PAA.

2.4 Process Control Strategies

Equipment specifications should consider process control requirements. Chapter 11 presents details of process control.

3.0 DESIGN

Following completion of bench and pilot testing and conceptual design, the preliminary design phase begins. The goal of this phase of the project is to advance the design from 10% completion to 30% completion. Key aspects of preliminary design include the following:

- Finalize the PAA storage and feed system design criteria;

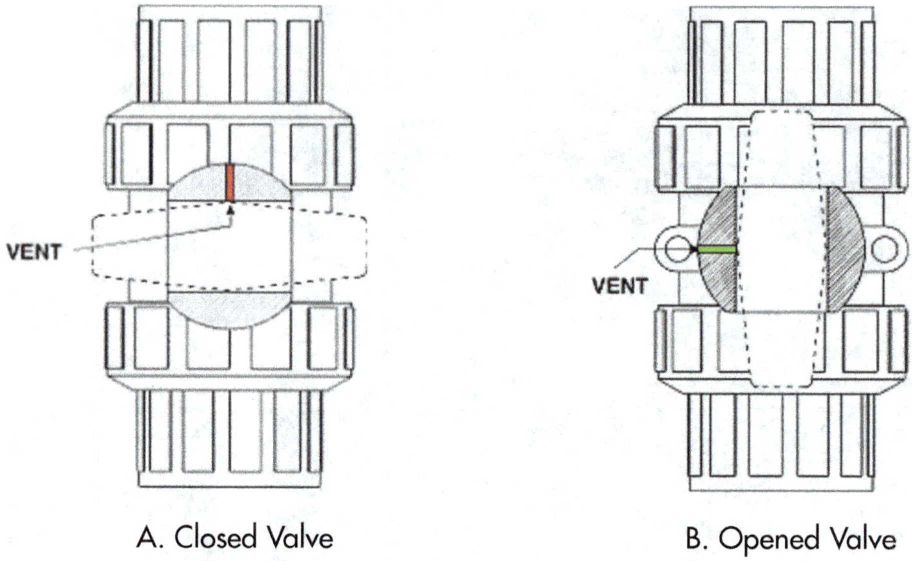

FIGURE 9.12 Vented ball valve (courtesy of Plast-O-Matic Valves).

- Develop PAA system procurement package, for leased systems only;
- Finalize PAA storage and feed system specification;
- Advancement of the design elements;
- Process and instrumentation diagrams to 90%;
- Process mechanical design to 60%;
- All other engineering disciplines to 30%;
- Develop preliminary specification table of contents;
- Develop preliminary engineering report; and
- Technical review of documents.

While all these aspects are key, the first three listed are critical. The system design criteria should be based on the information collected during the study phase. As discussed further in Section 4.0 of this chapter, there are two main ways the utility can obtain the equipment for the PAA storage and feed system. If the equipment will be leased, a procurement package should be used in the preliminary design phase. This allows the rest of the project to be designed around the system that will be used. If the utility will own the equipment, a procurement package is not necessary. The PAA storage and feed system specification is at the heart of the PAA system design because it will outline the technical requirements of the system.

The preliminary engineering report (PER) is used to document the project decisions made during preliminary design. The PER should include the following components:

- Introduction,
- Basis of design/design criteria,
- Description of PAA system,
- Supporting engineering disciplines,
- Cost,
- Schedule,
- Drawings, and
- Select technical specifications.

3.1 Final Design

Once the PER has been reviewed and approved, and, if applicable, the leased PAA storage and feed system has been procured, the project moves into the final design phase. The goal of the final design phase is to generate a set of construction-ready drawings and specifications that can be bid on by contractors. If a leased PAA storage and feed system is used, it serves as the basis for the final design; otherwise, the system shown in the PER will continue to be used during this phase of design. If leased equipment is used on the project, it is important to have approved shop drawings of the PAA system from the PAA system provider before advancing the project into final design. This will likely cause a delay between the conclusion of preliminary design and the start of final design.

The final design phase should include project milestones, typically 60%, 90%, and 100%. It is good practice to have a technical review of the 60% design package and to have an engineering discipline cross-check of the 90% design package. The project schedule and construction cost estimate should also be included at the 60%, 90%, and 100% milestones.

4.0 PROCUREMENT

This section focuses on the procurement of the storage and feed system equipment necessary for using PAA. Any structural, civil, and electrical components of the overall project will be procured using one of the project execution methods mentioned previously.

Water resource recovery facilities across the country have entered into PAA chemical supply agreements that range from "chemical only" annual contracts to "full service", multiyear agreements with custom design equipment and multiple services. Contrary to conventional wastewater treatment chemicals, PAA has attracted direct participation of chemical manufacturers offering full-service solutions that, in addition to chemical supply, include storage and feed equipment (leased or purchased), equipment installation, equipment commissioning and startup, operator training, and operation and maintenance services. This section discusses alternatives to procuring each of these items.

The design of PAA storage and feed systems must incorporate features that ensure safe and reliable operation, as discussed in Section 3.0. Given that PAA providers have ample experience in sizing, selecting, and installing PAA tanks, pumps, piping, and related pressure-relief devices, many utilities have opted for storage and feed systems designed and installed by the PAA providers, keeping responsibility for sizing, selection, and documentation on the PAA provider scope. Furthermore, by leasing equipment from a PAA provider, many utilities have minimized capital expenses associated with the conversion to PAA disinfection and have typically kept the responsibility of maintaining equipment on the PAA provider's scope. The following are examples of different equipment procurement approaches:

- In 2011, a utility in Florida purchased a feed system designed and installed by the PAA provider.
- In 2014, a utility in Texas entered into a 5-year lease of a bulk storage and feed system designed and supplied by the PAA provider, but installed by a general contractor. The lease can be extended according to the contract.
- In 2016, a utility in Kentucky designed and procured its own PAA storage and feed system.
- In 2017, a utility in Tennessee entered into a 15-year lease of a bulk storage and feed system designed, delivered, and installed by the PAA provider.

Regardless of whether the equipment is supplied by a PAA provider or a third party, owners can choose to use a general contractor familiar with the facility to install the equipment or to keep the responsibility of installation with the PAA provider. If choosing a general contractor for equipment installation, the owner should require close coordination between the general contractor, the engineer, and the PAA provider to avoid conflicts and delays. Furthermore, the owner should require the PAA provider to supervise unloading, storage, and handling of delivered materials to avoid delays caused by lost or damaged components. The PAA provider must also supervise and approve proper installation of all equipment before commissioning and startup. When considering using the PAA provider for equipment installation, owners should carefully evaluate the experience of each provider in installing equipment at a WRRF and request references from other entities operating equipment of similar size and complexity.

When equipment is supplied by the PAA provider (whether leased or purchased), the owner should strongly consider including the responsibility for functional testing of each component, system commissioning, and system startup in the scope of the PAA provider. Otherwise, the general contractor responsible for installation should be required to coordinate testing and system startup to the owner's satisfaction.

Peracetic acid providers typically include the cost of equipment maintenance (parts and labor for normal wear and tear items) in the equipment lease fee; however, that requirement needs to be included in the contract documents. Some municipalities have opted for a more comprehensive approach, giving the PAA provider responsibility for continuous operation of the system under a performance agreement. This approach may be desired when complex control strategies are required to maintain compliance. Table 9.1 compares equipment responsibility matrices for four different installations in the United States.

When considering the use of PAA disinfection, owners should evaluate the type of agreement that can be most beneficial for the WRRF in question.

TABLE 9.1 Equipment responsibility comparison.

State of installation	Design	Supply	Installation	Startup	Operation	Maintenance
FL	PAA provider	PAA provider (purchase)	PAA provider	PAA provider	Owner	Owner
TX	PAA provider	PAA provider (lease)	Owner's contractor	PAA provider	Owner	PAA provider
KY	Owner	Vendor (purchase)	Owner's contractor	Owner	Owner	Owner
TN	PAA provider	PAA provider (lease)	PAA provider	PAA provider	PAA provider	PAA provider

The evaluation of procurement alternatives should include, at a minimum, financial and operational impacts.

- Financial impact—Evaluation should include total capital cost versus annual operation cost. Equipment lease options reduce capital, but increase operating cost.
- Operational impact—Evaluation should include quantity and experience of operations staff available to operate and maintain the PAA disinfection system. Leasing equipment puts the burden of keeping the system operational on the PAA provider.

4.1 Alternative 1—Lease Equipment/Purchase Chemical

The procurement process should be structured to allow pre-selection of a PAA provider during design and before construction for this alternative. Once the PAA provider has been selected, design work will be coordinated with the PAA provider to integrate the PAA provider system with the owner's requirements.

The impact to schedule should be evaluated on a case-by-case basis, but owners should provide sufficient time on the front end of the project to evaluate and pre-select PAA providers. The PAA procurement decision should be made in the pre-planning stage of the project. If the lease option is selected, procurement should start as soon as the design criteria are established. Once the PAA provider is chosen, the detailed design can be completed to incorporate input from them. This input may result in a shorter construction schedule and fewer design changes. Consolidating responsibility on the PAA provider can simplify coordination and enable the owner to manage the schedule more effectively because a single point of contact will be responsible for design, installation, commissioning, and startup.

By leasing equipment, owners transfer capital cost into annual operating cost. When reducing capital expenses is desired, leasing equipment is recommended. Equipment lease fees can include installation, commissioning, and startup, further reducing capital expenses. Leasing fees are typically not paid until the system passes the acceptance test, so large projects could have a significant impact on cash flow.

4.2 Alternative 2—Own Equipment/Purchase Chemical

This alternative is a more traditional project delivery method in which the owner either designs or retains an engineer to design a system for procuring equipment and hires a contractor to install the equipment. A separate procurement would still be held to purchase the PAA. Even in this scenario, the

PAA provider will require a thorough safety review of the designed system and installation. The PAA provider will also ensure that all safety considerations are included and installed in the system before delivering any chemical.

5.0 REFERENCES

International Code Council. (2012). *International Building Code.*

International Code Council. (2012). *International Fire Code.*

International Code Council. (2012). *International Mechanical Code.*

Metcalf & Eddy, Inc. (2003). *Wastewater engineering: treatment and reuse.* Boston, MA: McGraw-Hill.

National Fire Protection Association. (2018). *Standard for Flammable and Combustible Liquids* (NFPA Standard No. 30). Quincy, MA: National Fire Protection Association.

10

Construction, Installation, and Commissioning

Sarah Stewart, Carl Hennagir, Joanne MacDougall, Samuel Lowen, Alberto Garibi, & Leonard Levine

1.0 INTRODUCTION

This chapter outlines critical tasks specific to PAA projects during each stage of construction, commissioning, and startup that help facilitate project success. Figure 10.1 lists these stages. In addition, this chapter identifies the construction team and roles; highlights the unique features of construction associated with PAA systems; reviews quality control; and discusses commissioning, startup, and optimization processes. While most of the implementation occurs in Stage 2, each of the stages is critical for successful project delivery.

1.1 Stage 1—Pre-Construction

Pre-construction is typically a short step in which rules are established with respect to communication, roles and responsibilities, document control, safety, quality, and schedule. Critical to this stage is developing a sense of team with all stakeholders, defining the project mission, and developing the critical success factors to achieve that mission. A pre-construction meeting is typically held at the outset of construction, with all stakeholders present.

1.2 Stage 2—Construction

Most of the work occurs here. The project goes from design to implementation through steps such as procuring equipment, demolition, construction, wiring, and programming—and all while planning for critical tie-ins, maintaining operation of the existing facilities, and planning for the next stages. Regular progress meetings are typically held during this stage with a frequency agreed to by the stakeholders, which is a minimum of at least once monthly, but possibly more frequent as construction commences.

1.3 Stages 3 and 4—Commissioning and Startup

Commissioning and startup are two separate phases, but are often grouped together. Section 2.0 of this chapter describes these stages in detail.

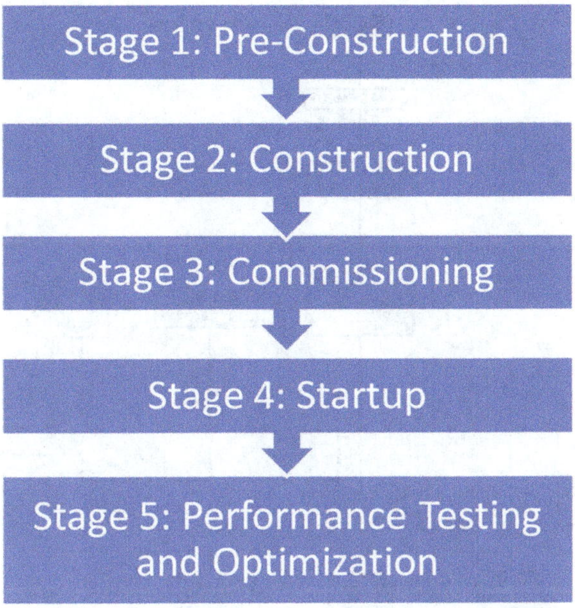

FIGURE 10.1 Stages of construction.

1.4 Stage 5—Performance Testing and Operation

This stage may run for a short or long amount of time, depending on the owner and the variability of the system, as will be described in Section 3.0. Process validation is extended sampling over a period of time to verify that the desired efficacy of the PAA system is achieved and within the dose range expected under varying conditions at the facility. This stage could be a matter of weeks, to months, depending on the amount of data to be collected and the performance of the system.

1.5 Construction Team and Roles

The construction team is made up of many different people and roles on a project and can vary depending on delivery method. Figure 10.2 shows a sample organizational chart detailing the various individuals and how interactions will occur.

1.5.1 Role of the Owner

The term, *owner*, in this section is intended to include not only the owner of the facility, but any party that represents the owner, such as facility operations and maintenance (O&M) staff, the owner's construction oversight personnel, project manager, and technical staff. The owner is ultimately responsible for the success of the project and the operation of the facility.

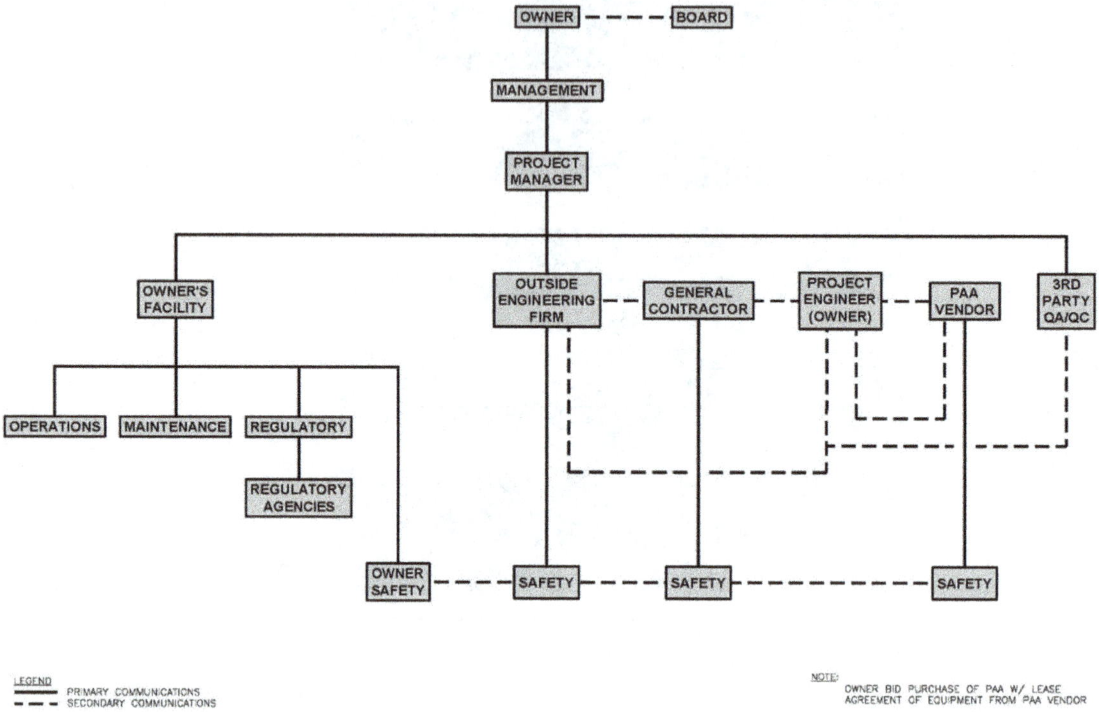

FIGURE 10.2 Construction organizational chart.

To that end, there are several aspects to the owner's role in PAA projects that are important as the project progresses.

1.5.1.1 Maintaining Compliance

It is critical that contract documents make it clear that the owner is responsible for permit compliance and that the contractor is responsible for conducting construction activities so as to not jeopardize this responsibility. The owner is responsible for coordinating with the various regulatory agencies before and throughout construction to keep them abreast of progress and informed of any planned shutdowns. The owner must communicate with designers to ensure that the construction documents include specific language limiting equipment downtime and require written owner-approved plans for all temporary outages and tie-ins to existing equipment. The contract documents must also make it clear which areas different operations groups require access to during construction (both vehicular and pedestrian) for operations, sampling, maintenance, and so on. If these areas are to be shared with the construction team, clear procedures must be developed to coordinate activities.

It is strongly recommended that planning for the key tie-ins required for construction be undertaken early in the design phase of the project so that any potential permit effects can be identified and reviewed with permitting agencies to gain required approvals before construction.

1.5.1.2 Coordination Between Stakeholders

Depending on the delivery method, the owner may hold separate contracts with a number of entities, including the design engineer, general contractor, and PAA provider. As such, it is the owner's responsibility to coordinate construction activities between all parties to ensure proper operation of the facility. It is important to evaluate how coordination will be implemented during design so that key elements can be included in the bid documents to avoid potential disputes during construction. The following are typical parties that may be involved in a project:

- Existing facility staff including management, O&M, and compliance groups
- Regulatory agencies
- Design engineer
- The general contractor and their subcontractors
- The PAA provider and their subcontractors
- The quality assurance/quality control (QA/QC) firm overseeing the proper installation of the system in accordance with the plans and specifications, if applicable
- Other contractors or vendors to the owner, such as other chemical suppliers
- Governing body, such as the city council or board of directors and other end users

Examples of typical collaboration events between stakeholders and their purposes include the following:

- Pre-construction meeting—The purpose of this meeting is to review the overall schedule, define roles and responsibilities, identify critical submittals and reviews, and establish the communication protocols for paperwork as well as verbal and written communication within the team. Attendees typically include all stakeholders.
- Progress meetings—These meetings are held frequently during construction. Often, the frequency depends on the preferences of the team or stage of construction and can vary from once a month to once a week. These meetings typically allow for updates on construction progress that include timing for critical reviews and quality checks, schedule status, and updates; they also often allow for discussion of

maintenance of facility operations plans as well as startup and commissioning plans. Quality and safety issues are also included in the agenda. Attendees typically include the owner, quality control/quality assurance personnel, the general contractor, the design engineer, and the PAA provider and their subcontractors.

- Pre-commissioning and startup meeting—These meetings review the commissioning and startup plan and roles and responsibilities. Additional staff may be needed for this testing, refining the testing scope and expectations for a complete commissioning as well as a complete startup. Attendees typically include all stakeholders.

1.5.1.3 Final Acceptance

The owner is ultimately responsible for final acceptance of the completed system, but may rely on a third party to recommend final acceptance (e.g., design engineer). It is important that a testing plan be developed by the engineer for new equipment before acceptance by the owner. Typically, these requirements are included in the bid documents with the provisions that require turnover of all as-builts, manuals, and other key supporting documents before final acceptance. An effective final acceptance testing plan should include key performance parameter metrics that can be accurately measured during system startup as well as the method that will be used to collect and process field data to determine whether the system meets performance requirements.

It is recommended that if the PAA provider either sells or leases the equipment that a representative of the PAA provider be involved in construction oversight during equipment installation, provide a certificate of proper installation, be responsible for proper commissioning with owner oversight, and take part in startup and testing as a support role to the owner. If the project includes a retrofit of existing facilities, the plan should consider the importance of continued operation of the existing disinfection system during the turnover.

1.5.1.4 Safety and Training

The PAA provider must be required to provide training to the owner on how to properly and safely handle the PAA and run the equipment supplied. This should include operations procedures and procedures for maintenance, if that is how the contract is structured. Procedures must include a list of the personal protective equipment (PPE) required for each operation or maintenance task.

If the PAA provider is to provide preventative and ongoing maintenance in accordance with the contract, the provider must have a reliable local workforce and supply of critical maintenance items locked up in a secure area at the facility. It is recommended that the facility personnel have access to the spares and be trained on performing maintenance on the equipment to cover an emergency where the PAA provider cannot get to the equipment in a timely manner.

The general contractor, PAA provider, and owner must ensure that the construction of the project is performed in a safe manner in compliance with all entities' safety programs.

1.5.1.5 Construction Oversight

This task can be provided by various entities, depending on the delivery method used (e.g., design/bid/build, construction manager at risk, design/build, or other methods). It is recommended that the PAA provider be responsible for construction oversight and signoff of their system, especially if they were responsible for the overall design and will be responsible for the system maintenance. It is also prudent that the overall design engineer and general contractor (depending on the contract type) be involved in construction oversight. The owner and owner's representatives, including the QA/QC subcontractor, must be involved in the oversight to make sure that the owner gets what was designed and specified.

1.5.2 Role of the Peracetic Acid Provider

The PAA provider's role varies, often based on the owner's preference. If the PAA provider is only selling chemicals and the owner is taking the responsibility to design and construct a storage and feed system that they own, then the PAA provider should be required to do reviews of the design and construction to ensure chemical compatibility of the PAA feed and storage equipment with PAA and ensure compatibility with the PAA delivery equipment/hose connections. If a lease/purchase arrangement is selected by the owner during procurement (see Chapter 9), then the PAA provider's involvement is much more in depth.

1.5.2.1 Construction by Peracetic Acid Provider

If the responsibility of equipment installation has been assigned to the PAA provider, close coordination with the owner and design engineer after preselection is required to define the construction phase schedule, procedures, plans, and reports necessary to ensure successful execution of the project. At a minimum, the PAA provider should prepare the following construction-phase documents for owner approval:

- Preliminary construction schedule—This includes key milestones and tasks. The schedule should include sufficient detail to facilitate coordination of contractor activities with the owner;
- Equipment installation and testing—This includes items such as hydrostatic testing of tanks and piping and electrical point-to-point verifications of instrumentation and controls;
- Site-use plans—Defining the laydown area requirements, including location of office trailers, storage trailers and material laydown, utility requirements, security and fencing requirements, and parking requirements;
- Equipment unloading, storage, and handling plan;
- Construction means and methods work plan(s);
- Construction safety plan;
- Maintenance of traffic plans;
- Outage requests—These would define when and how long outages would occur during construction that could apply to electrical or flow outages;
- Proposed testing procedures; and
- Equipment passivation plan—Define the types and amounts of chemicals used at the site for passivation and for disposal of chemicals from the site. The spent passivation chemical, typically nitric acid, is generally considered a hazardous material; therefore, a hazardous waste generator number is needed for a specialty company to retrieve the chemical and dispose of it.

1.5.2.2 Construction by Third Party

If the responsibility of equipment installation has been assigned to a third-party contractor, the PAA provider shall work closely with the owner and design engineer to review construction plans and procedures submitted by the contractor to certify that the equipment meets the PAA provider's minimum safety standards.

The PAA provider should furnish installation inspection services with the PAA provider's certified field technicians that, in coordination with the owner and engineer, will assist in validating the correct installation of equipment before commissioning and startup.

1.5.3 Other Stakeholders

Other stakeholders during construction include the design engineer and regulatory agencies.

1.5.3.1 Role of the Design Engineer

The role of the design engineer varies based on the owner's preferences and the scope of the design. If the owner hires an engineering firm to design the storage and feed systems, then the design engineer will be responsible for the overall coordination and successful implementation of the system, and verify that the project has been constructed as designed and has successfully reached the substantial and final completion milestones.

In a lease/purchase environment, the design engineer may be responsible for portions of the design, such as storage pads, road and electrical improvements to facilitate operation of a PAA system, and modifications and/or construction of a contact tank. However, the design engineer will not necessarily be responsible for the detailed design of the chemical feed and storage systems themselves because they are designed and provided by the PAA provider. Performance testing requirements for final acceptance of the project would also be outlined by the design engineer.

1.5.3.2 Role of the Regulating Agency

The regulatory agency is typically a stakeholder in the construction process that has a vested interest in the success of the project. They may have additional requirements during construction, startup, and validation testing that must be incorporated to the process to deliver a successful outcome for the project. Because the roles of the regulating agency vary greatly, early coordination to understand their requirements is a necessary step in developing a successful project.

1.6 Unique Features of Peracetic Acid Construction

Unique features of PAA construction include coordination between all parties to identify the scope and responsibilities of the various entities. Each of the parties working together is necessary for successful construction. It must be made certain that compatible materials are used and installed in the correct locations. If possible, the PAA materials should be stored in a separate area than the other project materials to prevent inadvertent use of incompatible materials (of particular concern is accidental use of incompatible gasket material). A clear understanding and designation of roles and responsibilities for various portions of the project must also be developed.

1.6.1 Passivation

Passivation of stainless steel for PAA systems is typically performed in the field. This is not a standard construction method for municipal applications and requires both the shipment and proper removal of passivating chemicals used on-site. In general, the responsibility for passivation of the system should be part of the PAA provider's scope. Owners may require PAA providers to submit passivation procedures as part of their proposal to ensure passivation methods and requirements are well understood by all stakeholders.

1.6.2 Coordination Between Third-Party Vendor and Contractor

Pre-selection of the PAA provider during design is also helpful in that if the PAA provider is supplying equipment for a general contractor to install, then approved equipment shop drawings, instrumentation and control diagrams, electrical wiring diagrams, and installation requirements can be defined before retaining a general contractor. The PAA provider should provide oversight of the construction and installation of their equipment to confirm that the installation meets all parties' requirements. If the PAA tanks and equipment are supplied by the PAA provider, they should certify that the PAA equipment is installed properly before commissioning and startup.

1.7 Quality Control

Quality control procedures during the construction phase should include detailed inspection and testing procedures that cover the following:

- Validation that equipment and materials delivered are unloaded, handled, and stored per vendor requirements and procedures;
- Validation that equipment and materials delivered match specifications and approved shop drawings, including materials of construction of gaskets, o-rings, and smaller components;
- Documentation that each component was physically inspected to ensure proper materials of construction will be installed;
- Validation that transfer of equipment and materials to the contractor for installation is well documented to avoid material loss; and
- Material status reports (MSR) to track inspection, validation, and transfer of materials during the construction phase. The MSR should be updated daily and made available to the owner and other relevant parties.

2.0 COMMISSIONING AND STARTUP

Commissioning is the cleaning, check out, and testing of all equipment to ensure the system is ready for startup and normal operation. This includes, for example, hydrostatic testing of tanks, piping, and fittings; calibrating instruments; checking control logic; verifying shutdown interlocks; bumping motors; and verifying valve operation. This work should be done with clean water and performed by the general contractor with the PAA provider, owner, and design engineer, depending on the process chosen, witnessing and signing off on proper installation. If the new PAA system is installed using existing infrastructure, commissioning and startup may need to occur simultaneously.

Commissioning work is often intertwined with ongoing construction work. For instance, hydro-testing may need to happen before all instrumentation is installed. This means the development of a complete commissioning plan and schedule is required to make sure all equipment will be tested, as needed, and all required personnel are available to complete commissioning in a timely manner at the proper point in the construction process.

In the startup phase, process water and chemicals are introduced to the system and normal operation is initiated. Finally, optimization of operating parameters and control system settings is necessary to document final operating parameters in the system's standard operating procedures.

During these stages of construction, document management and training become critical items between the contractor, PAA provider, and owner. Before commissioning and startup, the staff need to be trained on the operation, troubleshooting, sampling, and safety procedures related to the new system. The contractor and PAA provider need to submit draft and final O&M manuals, site-specific safety manuals, all relevant spare parts and spare parts lists, and other documents as required in the contract to the owner before starting up the system. Even if O&M of the PAA system is turned over to a third party, it is recommended that facility operations staff also have access to the equipment manuals and spare parts in the event of an emergency.

2.1 Commissioning

A commissioning plan would preferably be developed with the construction sequence plan during design. This is especially critical in a project where there is an owner, general contractor, and PAA provider all involved to make sure all required testing can be performed, adequate time is allotted for commissioning, and the proper order of construction is maintained. A critical component in commissioning is identifying which equipment needs water for checkout (such as pumps, mixers, or contact tanks) and which equipment needs PAA to commission (such as an online PAA analyzer or mass flowmeters). Permanent power supply is required before commissioning.

The commissioning effort typically involves running the equipment using clean water to verify it functions and meets the technical specifications before it is put into full service with chemicals. A tote of clean water or a water truck can be brought in to represent a tanker of chemical being delivered. Table 10.1 presents a simplified version of a commissioning plan.

2.2 Startup

A startup plan also needs to be developed with input from the owner, PAA provider, and general contractor to ensure activities are completed in the correct order. Designating a single point of contact to communicate and oversee all parties during startup can help to provide a smooth startup.

TABLE 10.1 Simplified commissioning plan.

Step	Item	Complete (Y/N)	Notes
1	Sign off on mechanical installation.		
2	Commission and startup electrical motor control center (MCC).		
3	Offload water into storage tanks (commission offloading pumps, floats).		
4	Fill contact basin with water.		
5	Run mixers (if needed) and confirm programmable logic controller controls.		
6	Run pumps with water to check function.		
7	Calibrate and verify flowmeters (if needed).		
8	Verify all interlocks.		
9	Drain water from PAA system.		
10	Remove floats, flowmeters, and anything else that cannot come in contact with passivation fluid.		
11	Passivate.		
12	Re-install floats, flowmeters, and so on to prepare for startup.		

Planning for startup should be conducted in concert with commissioning because they are often interlinked. Table 10.2 presents a sample startup checklist.

Startup can be a quick process, but once the system is on, it is fully functioning. Beyond the physical completion of construction, there are other aspects that include sampling, process controls, and monitoring/sampling to confirm the system is up and successfully running.

2.2.1 Sampling and Control

The sampling that needs to occur during startup can be performed by the owner, but it is typically performed by the PAA provider as a part of their contract. Typically, during the study and design phase, an initial process control scenario is developed that identifies dose ranges, control strategy, contact time, etc. Before startup, water quality sampling may begin to confirm that the water quality parameters that occurred during the initial testing (bench and pilot) are still occurring. In addition, owner coordination with the laboratory is needed to confirm that they have enough equipment on hand to start performing increased testing for PAA.

2.2.1.1 Initial Dosing

The process control and dosing strategy established during piloting and testing is normally used as a basis to determine operating parameters during startup. Often, it is common to overdose slightly at the outset until verification of system performance has been achieved. Then, careful observation and testing is conducted as the dose is optimized for long-term use.

2.2.1.2 Converting to the New Peracetic Acid Disinfection Strategy

Certain chemicals may react with PAA. While there is evidence that PAA and ultraviolet (UV) irradiation are mutually beneficial to one another, PAA and chlorine are not. Chlorine present in the water creates a PAA demand and thereby reduces the effectiveness of PAA. Further, with a chlorinated system, sodium bisulfite will remove both chlorine and PAA.

When converting from a chlorine-based (sodium hypochlorite or gaseous chlorine) disinfection strategy to a PAA strategy, both systems need to stay active, and the chlorine-based disinfectant will stop being fed immediately upon the introduction of PAA. The chlorine-quenching chemical remains being fed until the chlorine has had an opportunity to fully leave the contact tank before removing the quenching agent from the system. Note that, depending on the jurisdiction, the design and operating conditions such as effluent PAA concentration, the PAA label, and potential biomonitoring

TABLE 10.2 Sample startup checklist.

Item #	Task	Responsible party	Status
1	Verify all chemicals are in storage/feed tanks ready to go and the PAA basin is full of water.	Owner Constructor	
2	Programmable logic controller and instruments commissioned.	Engineering	
3	Sampler is installed and ready.	Constructor	
4	Piping inspected for miscellaneous blinds.	Constructor	
5	PAA mixers installed and operational.	Owner	
6	Operator training complete.	Supplier	
7a	Constructor begins removing pipe plugs.	Constructor	
7b	Design engineer directs operations to begin filling pipes and manholes with facility effluent.	Engineering	
8	Operations starts filling manholes and pipes with facility water.	Operations	
9	Stop filling and tell construction contractor when the level gets to bottom of the effluent line.	Operations	
10	Constructor can now pull isolation blinds.	Constructor	
11	Constructor confirms all blinds are pulled and turns over pipes and manholes.	Constructor	
12	Direct operations to resume filling.	Engineering	
13	Stop filling when the level is about equal to the level in the PAA basin.	Operations	
14	Open the valves at the inlet and exit of the PAA chamber.	Operations	
15	Operations inform engineer they are at this point.	Operations	
16	Facility signs off on transfer of custody.	Owner	
17	Engineer polls operations, constructor, and facility to ensure they are ready for startup.	Engineering	
18	Begin feeding PAA at desired flowrate.	Operations	

levels, it may be necessary to add a quenching agent to meet PAA discharge levels. It is also important to note that tests for PAA will also react with chlorine and show up as PAA; thus, monitoring PAA residuals will be inaccurate if residual chlorine is present. Additionally, if the discharge permit requires testing for chlorine in the effluent to confirm it is absent, it is important to note that PAA will measure as chlorine, so quenching may be required or the permit may require modification.

2.3 Safety

A safety plan specific for the startup/commissioning phase should be developed by the party responsible for commissioning, as described in Chapter 2. At a minimum, the safety plan should include the following:

- the PPE requirements for each commissioning task;
- a description of safety risks (e.g., fall, exposure, fire, etc.) associated with each commissioning task;
- emergency response procedures for each safety risk identified (the plan should include a list of hospitals and emergency locations where employees will be taken in case of emergency);
- safety data sheets for each chemical delivered to the site during commissioning;
- a training plan that ensures that all parties participating in the commissioning phase are familiar with the commissioning safety plan; and
- a procedure to validate that all required safety placards and signs are appropriately installed before commissioning, including
    - National Fire Protection Association (NFPA) safety diamonds on chemical storage tanks,
    - Signs illustrating what kind of PPE is required for the area, and
    - Piping flow direction and service decals.

The PAA provider, owner, and design engineer should approve the startup/commissioning safety plan before the start of the commissioning phase.

2.4 Training

It is recommended that the PAA provider instruct the owner's personnel on the proper safety procedures associated with storing and handling PAA and the safety requirements for operating and maintaining the equipment.

2.4.1 Equipment Operations and Maintenance

If the owner is going to operate and maintain the PAA system, then, before startup and commissioning, training sessions for the owner's O&M staff should be conducted so that everyone is familiar with the operation and maintenance of the equipment and dissemination of the approved O&M manual to the O&M staff. The training sessions should include a review of each section of the manual to provide O&M staff familiarity with the manual's contents. A review of the commissioning plan including sequence of inspections and tests, checklists, and other documentation to be used during commissioning should be included.

At least 2 weeks before the scheduled training, a detailed lesson plan, representative of the material to be covered during instruction, should be submitted to the design engineer for approval. Lesson plans shall consist of in-depth outlines of the training material, including a table of contents, resumé of the instructor, materials to be covered, startup procedures, maintenance requirements, safety considerations, and shutdown procedures.

The training plan should also provide a clear understanding of the system controls and what to do to address high flows, leaks, power loss, varying disinfection residuals, emergency stop procedures, etc. The operations staff need to understand not only "normal" operating conditions, but also emergency or atypical conditions.

If the PAA provider has a contract for O&M services for the owner, the owner's O&M staff should be given general training about the PAA system and their responsibilities, if any, including who they need to contact should they observe an issue.

2.5 Performance Testing

Validation, or performance testing, is typically performed after system startup to verify the system performance for a period of at least 30 days, although the duration can be adjusted to meet the owner's needs, to monitor the dosing points, PAA residual concentrations, chemical feed strategy, and effectiveness of PAA in disinfecting the wastewater. Typically, the validation testing plan is developed during design, whether it is by the design engineer as part of the performance specifications or by the PAA provider as part of their confirmation of system operation accuracy. The owner should be engaged in the development of the validation testing plan.

Depending on the contract language, either the owner or the PAA provider are responsible for collecting and analyzing the samples as well as distribution of data to the team. Developing a template for data input ahead of time can make confirming and updating design calculations a streamlined process for all.

As previously described, the dosing control strategy is typically selected during pilot testing and design. That strategy is then followed during the validation or performance test period, including gathering disinfection data (chemical dose, residual, influent and effluent bacterial counts) and also water quality and flow parameters such as turbidity, color, chemical oxygen demand (COD), and flow that define the conditions of the facility during the validation cycle.

2.6 Coordination

Typically, calls will be held weekly to evaluate the results and discuss whether changes need to be made to the dose or control strategy based on the actual results. Initial doses may be slightly higher than the design to confirm adequate efficacy, and then slowly lowered with time to optimize dosage while minimizing the risk of noncompliance.

If the PAA provider is performing the validation testing, collecting split samples and involving the owner in data collection as a secondary check may also be prudent (and vice versa if the PAA provider is performing the validation testing).

2.7 Acceptance Testing and Warranty Enforcement

Successful completion of validation testing will typically allow for either substantial or final completion to occur. The project will then enter the warranty period and long-term observation and optimization of the system will begin. Acceptance testing is the signoff by all parties that the system is performing as designed, both in a mechanical and physical aspect and by achieving the goal efficacy of the system.

Warranties are typically applied to cover equipment and physical construction to maintain a functioning system. A typical warranty duration is 12 months, but certain items may be covered for longer periods of time such as the storage tanks.

3.0 PROCESS OPTIMIZATION

After the permanent disinfection system is installed, a longer term (perhaps 3 months to 1 year) optimization period can occur. It is typically less involved than the validation testing period, but occurs over a longer period in an effort to collect performance data over varying seasonal conditions and influent and effluent conditions. The goal of optimization is to determine the most cost-effective dosing strategy, optimize operation of the system, and maintain permit compliance of the system.

3.1 Optimization Process

Similar to validation testing, both disinfection parameters and water quality and flow parameters are collected. The frequency may be reduced, but taken over a longer period of time. These data are often reviewed to see if there are trends where higher color or COD may necessitate higher doses of PAA to allow for a feed-forward strategy as opposed to maintaining an effluent residual. The placement of PAA analyzers to allow for a variety of feed controls and easy modifications during this period is a key feature for successful implementation. A design may also account for the ability to relocate them once the tank performance is ascertained during optimization. Peracetic acid online analyzers are a developing technology, and the selection of the right analyzer for the project is impacted by the water quality, temperature, and conditions of where it is to be installed.

3.2 Conducting Testing for Optimization

A template would typically be developed for this extended testing period for gathering data and to ensure that data are not collected just for collection's sake. The sample in Table 10.3 shows that multiple points throughout the tank are monitored as well as several influent flow characteristics. These data are collected over time and compared to the modelling data to refine the operation. The data collected during this period are often reviewed monthly for trends or to identify opportunities to fine tune the PAA dose. This step could result in long-term cost savings for the owner because it may allow for a decreased dose to achieve the same results.

TABLE 10.3 Sample optimization data collection template.

	Operator Data										Lab Data						
	Influent			PAA residual (ppm)							Entero				Influent		
Date	CB INF			PAA Analyzers		CBI SP1	CBI SP2	CBI SP3	CBI SP4	CBI SP5	CB INF	CBI SP1	CBI SP2	CBI SP5	CB INF		
	Flow	TOC	TSS	*Inlet*	*Outlet*	Point 1	Point 2	Point 3	Point 4	Point 5	Influent	Point 1	Point 2	Point 5	COD	Nitrite	Color
	MGD	ppm	ppm	ppm	ppm	ppm	ppm	ppm	ppm	ppm	ppm	ppm	ppm	ppm	ppm	ppm	
Start-up																	
Start-up																	
Week 1																	
Week 1																	
Week 1																	
Week 2																	
Week 2																	
Week 2																	

4.0 ANALYZING RESULTS

After the optimization period, a final design dose may be selected. Although overall monitoring should always be maintained of the system for permit compliance, there is a point of diminishing returns after which investment in additional sample analysis yields little to no further savings.

The point of diminishing returns is owner and site specific. There is a cost of time and effort for extensive sampling that may or may not allow for a reduction in chemical use and thus operational costs. Optimization of a PAA system is similar to the optimization of any process in that it has its limitations and, once satisfactory operation is achieved, it may not be worth the effort to continue extensive testing. There is also a safety factor, or additional dose, in selecting a dosing strategy that may need to be maintained, and that safety factor is dependent on the owner's risk tolerance.

11

Operation and Compliance Considerations

Kati Bell, Carl Hennagir, Ken Schnaars, Ashley Griffin, Vince Chiu, Theresa Gregonia, & Philip Harvey

1.0 INTRODUCTION

Generally, a well-designed and implemented peracetic acid (PAA) disinfection system is not maintenance intensive if operated in accordance with the operations and maintenance (O&M) requirements provided by the manufacturer and/or designer. Because each application is unique, the intent of this chapter is to provide principles and considerations that may be modified, depending on site-specific characteristics.

Depending on the water resource recovery facility (WRRF), the PAA feed may be operated in flow-paced, dose-paced, or other control modes, as described in this chapter. Because PAA is higher in cost on a unit basis compared to other liquid disinfectants, to be cost-competitive, typical PAA doses of 1 to 3 ppm for a typical 15- to 30-minute contact time are desirable. This concentration range is typical of secondary and tertiary municipal effluent when the bacteria reduction requirement is 2 to 3 log reduction. Each application, however, is unique, and bench and pilot testing are recommended before implementation, as described in Chapter 7.

Although the dose ranges referenced are typical for PAA to be cost-effective, there are examples, such as the City of Memphis, Tennessee, where the wastewater from the WRRF is high in color and there are other constituents that prevent use of other disinfection technologies. In the case of the City of Memphis, which receives high contributions of industrial discharges, PAA is dosed at higher rates than conventional municipal facilities to meet disinfection requirements. Whether PAA systems are large (like the Gulf Coast Authority Washburn Tunnel Facility in Pasadena, Texas; see Chapter 13 for a case study), or modest (like the Northwest Langley Wastewater Treatment Plant in Vancouver, Canada; see Chapter 13 for a case study), the same general operational principles apply.

Peracetic acid solution is typically applied neat (i.e., without carrier water). If carrier water is used, a shift in the equilibrium of the PAA solution can result, and, in some cases, may be significant enough to reduce efficacy. One of the most important considerations in PAA system operation is selection of the chemical injection, or dosing, point. Because PAA reacts quickly, an injection point that provides near-instantaneous mixing of the neat solution with the wastewater is ideal. Many chlorine system retrofits or conversions will have some form of mixing, whether it be static or mechanical mixers. In some applications, multiple feed points have been implemented to provide operational flexibility for low- and high-flow scenarios.

2.0 OPERATIONAL SAFETY

Peracetic acid is a strong oxidizer (slightly more aggressive than hydrogen peroxide, which makes up part of the PAA solution) and it breaks down into water, oxygen, and acetic acid upon decomposition. When product decomposition occurs in a closed system, typically resulting from a reaction with a contaminant, pressure can build, causing a safety hazard. Additionally, although the byproducts of decomposition are quite benign, in neat form, PAA is a highly reactive oxidant that can result in exothermal reactions.

Cleanliness is the primary consideration when working with PAA systems. Because PAA is highly reactive, it is important to prevent contact with light thermoplastics like acrylonitrile butadiene styrene (ABS) and nylon, dirt, dust, debris, and transition metals like iron, copper, and zinc. Good housekeeping and best management practices for cleaning tools, equipment, piping, and fittings that can contact PAA solution will prevent unwanted reactions. Contaminated water, such as water from a rusty fire hydrant, can be a source of PAA decomposition if it comes in contact with PAA; therefore, using clean, potable water for flushing prior to use with PAA is critical. Totes or drums should not be stored on wooden pallets because wood is organic and could catch fire. As such, it is important that vents on PAA storage vessels and piping systems are not blocked. Consult the chemical compatibility charts from the vendor for materials that are compatible with PAA solutions before use. In general, good chemical handling practices will allow safe use of PAA. This includes never returning unused PAA back to the original container (because this may return impurities). Chapter 2 provides additional information on safe handling of PAA.

2.1 Peracetic Acid System Hazards

The most significant process risk is over-pressurization of storage containers upon chemical decomposition. Because PAA releases oxygen upon decomposition, it is not recommended that PAA be stored indoors without mechanical ventilation and vapor monitoring equipment. Design considerations for a safe PAA system include provisions for venting to avoid trapping off gasses from PAA decomposition. The risk of combustion in a well-designed PAA system is low; Chapters 2 and 10 explain design considerations that address the major hazards of PAA.

Operationally, in addition to over-pressurization, the other significant safety risk is transport to the site and/or offloading at the site. If PAA is transported to a site in bulk, it is typically transferred from tanker to bulk storage. If totes (or larger bulk tanks) are used, transfer between totes or swapping totes/tanks is required. In either case, careful operation is the primary risk mitigation measure to minimize the potential for contamination and the need for worker contact and product handling. Special care should be taken to ensure that storage tank inlet(s) are kept clean, and inlets are capped between deliveries to avoid inadvertent contamination.

Contact with commercial PAA solutions used for wastewater disinfection can cause several safety issues, including burns and damage to eyes and skin. To protect staff from damage associated with exposure to PAA, precautions should be implemented related to chemical handling, as described in Chapter 2 and herein. Regular system maintenance should be performed with chemical systems offline, if possible; if maintenance needs to be performed while the system is operating, proper eye/face shields and skin protection must be used.

As with all other systems, an established lockout/tagout (LOGO) procedure must be published and followed. In the event of a spill, dilution of spilled PAA should be done with copious amounts of water, in a ratio of at least 10 parts water to 1 part PAA, to prevent combustion; spilled PAA solutions will have very low pH and may need to be neutralized with a material such as soda ash or a mild alkaline solution, and spilled material may need to be diked if there is concern related to release to a sensitive waterway.

Aside of chemical exposures and spills, like other disinfection systems, tripping and slipping hazards may exist around channels in the contact tank. This is particularly a risk when grates covering channels are removed; therefore, normal safety precautions should be taken when working around these open-channel structures.

While PAA can maintain its solution strength long after the expiration date, there are times when PAA disposal may be necessary. Peracetic acid disposal may fall under state and federal waste disposal rules; therefore, it is important to consult the manufacturer guidelines, the safety data sheet (SDS), and local and state requirements regarding proper handling procedures and proper disposal for spills. Released PAA must be neutralized, and this is typically conducted using nearby wash water supply to dilute the spill. The dilution of the neat solution causes decomposition of the chemical to harmless byproducts, including acetic acid and water. Operators must refer to the safety data sheets (SDS) provided by the PAA supplier for additional safety guidelines.

2.2 Electrical Hazards

To prevent electrical hazards, all safety and operational precautions required by the National Electric Code, Occupational Safety and Health

Administration (OSHA), local electric codes, and equipment systems manufacturers should be followed and include the following precautions:

- Proper grounding,
- Lockout/tagout procedures,
- Use of proper electrical insulators, and
- Installation of safety cutoff switches.

Proper grounding and insulation of electrical components are critical for protecting operators from electrical shock and protecting the equipment. To minimize electrical hazards, ground-fault-interruption circuitry should be provided with each module. Provisions enabling isolation and lockout for maintenance, related to chemical feed and mixing systems, should be included in the design. Control of all lockout systems should remain local.

3.0 PROCESS MONITORING AND CONTROL

Peracetic acid disinfection control systems use process instrumentation for monitoring, reporting, and control. This section describes the most common process instrumentation associated with PAA disinfection systems, potential uses for process control, and O&M factors. The section also provides guidance for disinfection system warnings and alarms.

3.1 Instrumentation

Similar to other chemical dosing systems, commercially available instrumentation is readily available to automate and monitor PAA systems. Other than verifying chemical compatibility of the wetted components, few other special considerations are required. Typical parameters that are monitored include flow, level, and temperature. When chemical compatibility is considered and implemented in design and construction of PAA facilities, there are no unique instruments required (other than a PAA vapor sensor).

The most basic form of chemical dose control is flow-pacing, with the desired dose set by the user and PAA pumped (dosed) at a rate proportional to facility flow. Other forms of control are available (e.g., feed-forward control) with the use of additional instrumentation, which detect (or proxy) residual PAA in liquid form. However, the user should consider the potential cost savings that may be achieved with the tighter control using online instrumentation versus the burden of instrumentation maintenance because residual PAA sensors can be labor intensive. Utilities may choose to dose PAA based on pump curves derived from drawdowns and/or use online flowmeters. The strategy of quantifying how much PAA is dosed depends on the desired level of detail of both the designer and user, but reliable flowmeters are available if instrumentation is desired.

In addition to flow, pressure indicators should be considered in storage vessels for both safety (overpressure) and for assessing normal operating conditions. Temperature is also a consideration for bulk storage (whether in totes or tanks), and temperature sensors provide a warning if unwanted accelerated oxidation is occurring. Instrumentation to determine the amount of PAA stored may assist in timing of deliveries, and this can be implemented using a level transmitter, pressure transmitter, or simple visual float. For containment, some instrumentation (e.g., level switches, conductivity probes) can be used to warn when secondary containment is activated. Floats or point-level switches can also be used in pump casings or skids to indicate pump leaks quickly.

3.1.1 Wastewater Flow and Contact Time

Flow instrumentation is typically not provided by PAA vendors, and data from a facility effluent flowmeter is typically provided as an input to the disinfection control system. The data may be relayed through the facility supervisory control and data acquisition (SCADA) system or, particularly at smaller facilities, there may be a direct data connection between the effluent flowmeter and the disinfection control system. Magnetic flowmeters and Parshall flumes with level instrumentation are the most common effluent flowmeter types. Facility operators should consider how the recorded effluent flow corresponds to the flow through the PAA disinfection system.

Flow at the contact tank may differ because of the addition or diversion of various recycle or service water streams, or there may be a time offset if the effluent flowmeter is not used at the PAA feed location. If the difference in actual flow and the flow signal is significant, or highly variable, a more accurate flow may be calculated in a programmable logic controller (PLC) by adding or subtracting other measured or estimated flows from the recorded effluent flow, with the calculation result sent by the facility SCADA to the disinfection control system. Operators should refer to the flowmeter manufacturer manual for maintenance and calibration requirements.

3.1.2 Chemical Feed Pumps, Pump Calibration, and Maintenance

In selecting the PAA pump, the main consideration is durability of the wetted parts. If peristaltic pumps are used for dosing, very little maintenance is required other than regular tube changes. There are various styles and types of peristaltic pumps on the market, and the main consideration is tube life because PAA is much more corrosive on pump tubing than sodium hypochlorite. Depending on the criticality of disinfection, a run-to-failure approach may not be prudent. Therefore, testing to know the actual service life of a tube would optimize periods between tube changes before failure. Using this approach, staff at the Northwest Langley Wastewater Treatment Plant have found that, for Norprene® chemical tubing, changes are required every 2 weeks of continuous running to stay ahead of tube failure. However, with other styles of peristaltic pumps, which use a tube within a disposal head filled with lubricating liquid, the tube life can be 6 weeks or greater.

It is important that PAA is accurately metered into the effluent because it is necessary to provide an adequate dose, but not an overdose. As with any chemical dosing applications, it is important to verify actual pump performance (e.g., using drawdowns), especially after peristaltic pump tubing changes. Because of the corrosive nature of PAA, wetted parts may wear sooner than with other chemicals. The durability of the wetted components varies greatly among pump suppliers; the range of operational time can be 350 to 5000-plus hours between failure, depending on the pump and application. Consultation with other users, PAA suppliers, engineers, and designers may provide some useful operational information on the latest pumping innovations and experience. Verification with testing is best to determine the actual service life of wetted components.

In all cases, chemical feed pumps must be calibrated periodically to ensure proper chemical feed rates. It is recommended that PAA feed pumps are calibrated monthly. To check dose rates of the PAA feed pumps, the operator must use a calibration column, unless an online chemical flowmeter has been provided in the system (and the flowmeters should be periodically calibrated). The calibration column is used to conduct verification of the feed rate by conducting the following steps:

1. Turn off the chemical feed pump.
2. Open the valve for the calibration column to fill up to the calibration column.
3. Fill the calibration column up to preselected starting point and record the starting reading.
4. Close the valve between the PAA storage tank and the calibration column.
5. Start the PAA chemical feed pump and operate it for a preset time.
6. After the preset removal time from the calibration column, stop the pump and record the final reading in the calibration column.
7. Use the following formula to determine the chemical pump feed rate:

$$\text{Chemical feed, mL} = \text{starting level, mL} - \text{finished level, mL} \quad (11.1)$$

$$\text{Chemical feed rate, mL/min} = \text{chemical used, mL/feed time, minutes} \qquad (11.2)$$

Regardless of the type of chemical metering pump, drawdowns should be performed periodically and after every diaphragm or tubing change out to verify the dosing curve. Care should be taken with tube changes and, accelerated by the corrosiveness of PAA, a pinched tube will fail quickly. For pumps that require more frequent tube changes, users may consider installing quick-connects and have the tubes pre-fitted to permit quick and safe tube changes, as shown in Figure 11.1. Some pump manufacturers have these quick-connects as options. In dosing mode, the system should be designed to quickly and safely isolate the pump to facilitate tube changes, which includes a shutoff valve to close both the suction and discharge with a safe bleed to relieve residual pressure.

4.0 STRATEGIES FOR PROCESS CONTROL

Although successful PAA dosing can be achieved by a simple flow-pacing regime, the time-dependent water quality demand on PAA at some WRRFs

FIGURE 11.1 Example quick release for maintenance (photo courtesy of Vince Chiu).

requires other process control schemes to optimize dosing. In many cases, online monitoring for various parameters is needed to control the process. While high-level information about online PAA residual analyzers has been provided throughout this manual, the following sections provide additional descriptions of other online water quality analyzers related to the various process control strategies.

4.1 Flow/Dose Pacing

When small doses of chemicals are fed to large flows, the simplest approach to controlling the process is to pace the chemical feed automatically. The pacing of the feed can be tied directly to the flowrate of the flow stream being treated. Typically, a 4 to 20 mA signal from the main effluent flow is sent to the local PAA feed pump PLC, which is configured to provide a 4 to 20 mA output that will drive a liquid mass flow controller to deliver chemical. This is the simplest process control method and, when effluent quality and PAA demands are consistent, is the most commonly used method for PAA disinfection system control.

4.2 Residual Feedback

The next level of automatic control that can be implemented uses online residual measurements. Changing wastewater characteristics can result in changes to the PAA demand, and that can be measured using residual monitoring methods, as described in Chapter 4. The PAA residual measurement is reported to a PLC, which reacts to eliminate differences between the measurement and the desired setpoint by sending a control signal to an actuator. The main drawbacks to this process are the lag time reflected in the location of the residual meter because it is typically minutes to hours downstream of the point of chemical application and, when flowrates are highly variable, there can be over- and undershooting of the feed rates. Dead band and hysteresis analytics can be applied to minimize oscillations, but this results in a more complex process control system.

Although more efficient control of PAA feed could be possible using a proportional-integral-derivative (PID) controller to maintain the process closer to setpoints and more closely match the responses of chemical demand, it is rarely implemented. When it is implemented, the configuration for PAA disinfection control system is achieved using a PID loop in a cascaded control system. The PAA controller calculates the chemical feed rate required based on the difference between the input PAA setpoint and online PAA residual measurement. The drawbacks to feedback control are that the response is implemented only after a process disturbance has occurred; tuning the system to match large and/or frequent disturbances is complex; and it is not possible if the response variable cannot be reliably measured, which is typically the challenge in wastewater disinfection applications.

4.3 Feed Forward

The concept of feed-forward process control is that proper dosing of PAA can be provided to achieve the target microbial reduction in a proactive mode, rather than the reactive mode that is used with residual control. The feed-forward dose control algorithm is typically based on a known relationship, as exemplified in the following equation:

$$PAA_{dose} = PAA_{setpoint} + PAA_{demand} \tag{11.3}$$

The equation for the PAA dose is then used as the dose for flow pacing, using the effluent flow signal as described in Section 4.1. Typically, an initial PAA residual is measured at the beginning of the contact tank, at approximately 1 minute from the injection point. This residual measurement is dependent on two factors: the PAA dose applied and the PAA demand associated with oxidation of nontarget compounds in the effluent flow. Flow also impacts the apparent PAA demand because of the time-dependent nature of the demand kinetics and PAA auto-decomposition described in Chapter 3. Additionally, flow impacts the hydraulics of the disinfection channel and, in some scenarios, can result in non-ideal flow for some flow regimes.

Peracetic acid demand is assessed through statistical analysis of PAA residual as a function of various wastewater quality parameters and flow. The best wastewater quality parameter for use in the feed-forward dose control algorithm must be selected based on site-specific testing and a statistical evaluation to identify the water quality parameter with the greatest sensitivity of the parameter to PAA dose (change in PAA dose/change in water quality parameter).

Based on experiences from testing at full-scale facilities such as in Memphis, Tennessee, there are water quality parameters that can be used for process control in a feed-forward control scheme. These parameters include wastewater color (as measured at 455 nm) and chemical oxygen demand (COD) (Block et al., 2015). It is also possible that ultraviolet transmittance (UVT) at 254 nm or oxidation–reduction potential (ORP) could be used; however, this has not yet been demonstrated at full scale as of the date this chapter was written.

Control systems for PAA disinfection can provide a variety of warnings, minor alarms, and significant and critical alarms. The terminology and categorization will vary by project-specific requirements. The warnings and alarms may be locally annunciated at the PAA system control panel(s) and/or they be transmitted to the facility SCADA system for remote annunciation and historical logging. For facilities that are not staffed around the clock, it is recommended that, at a minimum, significant/critical alarms are connected to a remote annunciation system (such as an auto-dialer) to notify off-site staff of the alarm.

5.0 OPERATOR CONSIDERATIONS

While PAA is not unlike other liquid chemicals used at WRRFs, there are some guidelines that may be useful to operations staff who have not had direct experience with PAA. Some of the key considerations are around chemical deliveries and daily activities associated with operations of the systems.

5.1 Chemical Storage and Delivery

Peracetic acid is delivered to WRRFs in various types of containers, and the chemical is stored at the facility in either tote-type containers or bulk storage tanks. Peracetic acid that is delivered to large facilities is typically trucked and then transferred to bulk storage tanks (see the case studies in Chapter 13 for photographs of bulk storage systems). The PAA storage tanks are typically sized to allow for a minimum of 30 days of storage capacity; Chapter 8 presents additional details on tank sizing. Bulk PAA is sometimes delivered in a truck that has several totes, which are transferred to bulk storage tanks; however, more recently, PAA can be delivered in bulk storage trucks and transferred to bulk storage tanks. If bulk storage tanks are used at a WRRF, they are typically plastic-type storage tanks. Peracetic acid manufacturers typically recommend that bulk plastic storage tanks be replaced approximately every 5 years; using tanks after their expected life could result in safety concerns such as damaged or leaky tanks.

A tote system may be installed at a small (<4 ML/d [<1 mgd]) or medium (4 to 38 ML/d [1 to 10 mgd]) WRRF. Individual PAA totes hold approximately 1140 L (300 gal) per tote. In some cases, several totes are required, and these totes must be stored properly, as described in Chapter 2. Because of the long shelf-life of PAA, in some instances even small facilities will install bulk tanks, which can accept a full truckload of chemical.

5.2 Daily Operations

It is important that operators make detailed observations when making daily rounds in the PAA chemical feed and storage areas. Operators must carefully observe and record chemical liquid levels, listen for any unusual noises, look for leaks, observe and record gauge readings, and monitor other operations. This section includes operational checklists (Table 11.1 and Table 11.2) to provide operators with an easy-to-follow and thorough form to use while making daily rounds so all operational aspects are checked. Peracetic acid systems are not complicated, but they do require operator attention. Whether it is a small system (Figure 11.2) or large PAA system, all PAA disinfection systems require operator attention.

Table 11.1 provides an operations checklist for small PAA feed systems that includes recommended activities for operators when they are making rounds and inspecting the PAA system. The checklist is only a recommendation and should be modified to meet the requirements of the WRRF where the PAA tote system is installed.

Just as a small PAA system requires operator attention, a large (>38 ML/d [>10 mgd]) PAA disinfection system, which may have more automation, also must be inspected daily to ensure that the system is operating properly. Table 11.2 presents a checklist that includes recommended activities for operators during rounds to ensure that the PAA system is functioning

FIGURE 11.2 Small temporary PAA feed system installed at the Nashville Metropolitan Water Services Dry Creek Wastewater Treatment Plant, Madison, Tennessee (photo courtesy of Kenneth Schnaars).

TABLE 11.1 Operator checklist for small PAA feed systems.

Item	Yes	No	Remarks
What is the facility effluent flow?			
What is the PAA residual reading?			
Are there any tank leaks?			
Is there adequate chemical in the storage totes?			
Do the chemical tote(s) need to be changed?			
Are the process valves set properly?			
Are there any leaks at the feed pumps? If yes, then which pump? • Pump 1 • Pump 2			
What is the pump output speed? • Pump 1 • Pump 2			
Are there any unusual noises coming from the pumps?			
Are there any housekeeping needs?			
Are there any maintenance needs?			

properly. The checklist is not necessarily a complete list of activities that may be needed for operations and, as such, it should be modified to meet the site-specific requirements of the WRRF where the PAA disinfection system is installed.

6.0 ELECTRICAL SYSTEM AND SUPERVISORY CONTROL AND DATA ACQUISITION CONSIDERATIONS

While PAA disinfection is a simple technology that has generally low electrical power requirements, reliable electric power is essential for successful disinfection operations. Further, many PAA disinfection systems are provided with control systems needed to monitor and adjust parameters related to dose and to warn facility operators of inadequate disinfection.

TABLE 11.2 Operator checklist for large PAA feed system.

Item	Yes	No	Remarks
What is the facility effluent flow?			
What is the PAA residual reading?			
Are there any bulk storage tank leaks?			
Is there adequate chemical in the bulk storage tanks?			
Are the truck unloading transfer pumps ready for operation and functioning properly?			
Are the process valves set properly?			
Are there any leaks at the feed pumps? If yes, then which pump? • Pump 1 • Pump 2			
What is the pump output speed? • Pump 1 • Pump 2			
Are there any unusual noises coming from the pumps?			
Is the bulk storage tank leak and level instrumentation and other PAA instrumentation operating properly?			
Are there any housekeeping needs?			
Are there any maintenance needs?			

6.1 Electrical Power Requirements, Reliability, and Redundancy

Some WRRFs have installed power meters at the motor control center to independently record incoming power quality and total power consumption of various parts of the treatment facility. Power quality data may be needed to identify and troubleshoot power quality issues with the electric utility if poor electrical quality results in interruptions of chemical feed system operations. To prevent electrical hazard, all safety and operational precautions required by the electric code should be followed. (Refer to Chapter 9 for information on electrical safety design considerations.)

Both power quality and power redundancy are important design considerations for any system to provide continuous disinfection. Chapter 9 includes a discussion of the design considerations for power quality and redundancy. It is also important to note that any electrical systems should be inspected and maintained regularly. The operations program should include periodic startup or testing of generators or other electrical equipment that is part of the electrical backup for the disinfection system.

6.2 Control Cabinets/Programmable Logic Controllers

Programmable logic controllers, or customized control boards, are typically incorporated to system control panels. Control boards/PLCs allow control systems to monitor the signals originating from online analyzers/monitors. The control boards or PLCs may also contain algorithms and equations for flow pacing, dose pacing, or other process control methods by controlling pump output.

The power services to PAA system components are generally straightforward. For example, 120-V air-conditioning (AC) service is generally required for instrumentation, such as online residual analyzers; unless the flow requirements for neat PAA are high, this service is generally adequate for the chemical feed system.

6.3 Supervisory Control and Data Acquisition System Integration

Peracetic acid disinfection systems may communicate with facility SCADA systems, particularly at larger installations. The communication will include PAA PLCs sending system status and alarm information to the SCADA system. Field instrumentation data from outside the scope of the PAA system, such as effluent flowmeters or residual analyzers, can be sent to the PLC via the facility SCADA system. The SCADA system can also be programmed to provide PAA PLC information regarding upstream process changes that influence influent to the PAA system, such as status of pump stations or filter backwash processes. In some instances, the facility SCADA system may be tasked with active control of the PAA disinfection system.

Peracetic acid system statuses, warnings, and alarms may be locally annunciated at the control center and/or be transmitted to the facility SCADA system for operation staff monitoring, historical logging, and remote annunciation. The SCADA integration can allow significant/critical alarms to be connected to a remote annunciation system (such as an auto-dialer) to notify off-site staff of the alarm. Some systems may be configured so that facility SCADA systems will automatically shut down PAA equipment upon certain critical alarms or conditions outside of the PAA system scope.

Facility SCADA systems may include historical logging and trending features. Data collected by the master PLCs, such as PAA tank status, elapsed time meters, warnings, and alarms, can be stored in the SCADA system or an external database. Trending and reporting features can be set up to fulfill the needs of operations staff. Typical parameters that are used for trending include PAA dose, PAA residual, tank volume, and other online data that may be used to evaluate system performance over time. Trending effluent flow and PAA dose and residual is strongly recommended on an ongoing basis because this information can be used to reassess capacity of the disinfection facilities at any point in time. Trending can also be used to estimate maintenance costs or validate warranty claims when process performance guarantees are provided as part of the system package.

Peracetic acid disinfection systems are typically provided with one control cabinet, unless the disinfection system is split. In systems with multiple contactors, the control boards/PLCs in each cabinet must communicate with each other to implement all but the simplest manual control schemes. Facility staff should always be aware of the system failure modes in case of loss of communication. For example, how will the PAA control system react to a loss of the effluent flowmeter signal, either because of instrument failure or a SCADA system failure? Failure modes must be programmed into each control cabinet for loss of all external data sources and documented in the O&M manual.

7.0 ANCILLARY SYSTEMS

Ancillary systems are critical to the storage and management of PAA and should be operated and maintained properly to prevent problems like corrosion, contamination, or service interruption. Without proper maintenance and operation, it may be possible to overfill tanks, contaminate product, or even lose containment in an emergency. These problems present safety, health, and environmental consequences and should be avoided by monitoring and verifying ancillary systems.

7.1 Chemical Transfer/Unloading Station

Unloading/transfer stations for receipt of chemical should be designed with appropriate materials by an experienced engineer with guidance from chemical suppliers. Safety interlocks are highly recommended and should be tested regularly. During transfer operations, multiple operators should be present, whether remotely while monitoring automation systems or on-site in appropriate personal protective equipment. Delivery drivers will likely be experienced and knowledgeable, but there needs to be a site representative present to monitor facility equipment and to act if need be.

If tank level devices are not in view of the unloading station, a site representative should be present to relay information to the driver. Hoses, camlock devices, and other product contact tools should be kept clean and stored neatly to prevent contamination of the product. All transfer lines should be marked with safety and Department of Transportation information to help prevent cross-contamination with other chemicals. Compressed air pressure and emergency equipment status should be verified before unloading.

During transfer, it is important to monitor temperature and watch for rapid heating because this may indicate rapid decomposition and a dangerous situation. Monitor pressure through lines during unloading as well as tank levels. To evacuate the system, use pump case drains or low point drains and regulate air pressure to 3.4 bar (50 psi) or less for blowdown. If residual PAA is not vented or may dry in long-term storage, rinse with deionized (DI) water or vent hoses.

Transfer pump maintenance should be performed per manufacturer suggestion and monitored for pressure and power draw to predict bearing or seal failure. Torque value should be checked at regular intervals on flange bolts and pipe supports. A hydro-test should be performed at regular intervals and after a line break through all pipes, permanent hoses, and gaskets. Calibrate and test instruments and safety interlocks at regular intervals. Operate backup systems in storage at regular intervals to ensure readiness.

7.2 Air Compressors/Blowers

Air compressors should be monitored for air pressure, flow, and humidity. Apply American Society for Testing and Materials (ASTM) D4285-83, *Standard Method for Indicating Oil or Water in Compressed Air*, at points of use and any manufacturer required maintenance at regular or prescribed intervals. Air compressors should be equipped with flow and pressure instruments and monitored during operation via SCADA or local indication.

Scrubbers should be regularly monitored via opacity, velocity, pH level, and density of liquor testing. Scrubber media should be monitored for any crystallization and rinsed with flush water regularly. Collection lines and low point drains should be inspected regularly. The liquor pump should be inspected or rebuilt per manufacturer specifications and equipped with an hour meter. The blower interior case should be inspected for thickness, and the impeller should be inspected for corrosion and balance.

7.3 Sump Pumps and Spill Management Systems

Sump pumps should be field and/or bench tested regularly and any manufacturer suggested maintenance should be followed. Piping and fittings should all be hydro-tested at regular intervals. Concrete should be inspected for containment. Spill kits should be inspected and stocked regularly and should be on-site always. Any rarely used pipelines should be tested at regular intervals.

Should a spill occur, HAZWOPER personnel are recommended to be on-site. Large spills (4 L [1 gal] or more) may require the use of a self-contained breathing apparatus and Class A or B suits. Spilled PAA will likely become contaminated and will decompose; therefore, dilution of the chemical is recommended as soon as possible. Neutralization of PAA can be achieved with material such as soda ash, sodium bisulfite, and sodium thiosulfate. Sand or dry earth can be useful with small spills, but must be neutralized before disposal. Neutralization for disposal will produce significant heat and should not be performed in plastic containers. All disposal must conform with hazardous waste guidelines. Neutralized material is likely nonhazardous.

8.0 TROUBLESHOOTING

Peracetic acid disinfection is effective when PAA reacts with microbes. A number of factors, including chemical dosing, contact time, temperature, pH, hydraulics, and the presence of inorganic and organic matter present in water, can affect PAA disinfection efficiencies. If there are issues with disinfection compliance, it is often related to one of the following factors.

8.1 Electrical/Instrumentation Issues

Even if the electrical system is designed properly, there can be issues associated with power failures and poor power quality that can interrupt the operations of the chemical feed pumps and analyzer systems. If these issues occur, backup power can be provided, particularly because these systems require such small amounts of power. If there are other issues with electrical failures, it could be associated with the wire connections, which should be checked on a regular basis.

8.2 Chemical Dosing

Similar to sodium hypochlorite, PAA can off-gas and produce bubbles in chemical feed lines that can cause air locking; a good design will accommodate this by providing off-gas valves at high points in the system. In addition to gas lock, chemical dosing can be impacted by the age of the PAA. Although PAA has a very long shelf life, if seasonal disinfection is conducted and there is carryover of unused chemical from one season to the next, chemical aging can occur, potentially reducing the solution strength. If there are no issues with calibration of the chemical feed pumps, and unusually low residuals are measured, it is advisable to conduct a chemical assay of the bulk chemical, as described in Chapter 4.

8.3 Hydraulics

To provide disinfection, PAA must be distributed and mixed throughout the entire flow. If there are issues with disinfection compliance at high or low flows, this is often indicative of short-circuiting or other blockages in the contact tank channels. Regular equipment and channel cleaning should be conducted to maintain the channel hydraulics as well as eliminate any potential issues with chemical demand by biofilms that can form, particularly at the end of the contact basin where PAA residuals may decay.

8.4 Upstream Process Impacts

One of the most significant considerations in the design of a PAA disinfection system is understanding the effect of water quality on both the efficacy and efficiency of PAA disinfection. Treated wastewater effluent includes both inorganics and organics. Recent research evaluating the impact of various constituents showed that inorganics and organics had different roles in PAA decay. In this study, inorganics were shown to affect the disinfection rate constant, which is what is responsible for PAA residual concentrations at a given contact time, and organics impacted the initial PAA demand, thereby impacting the minimum required PAA dose (Henao et al., 2018).

If the secondary treatment process experiences upset and biochemical oxygen demand (BOD) removal is poor, residual organics can impact the PAA demand; further, if inorganic coagulants are used for phosphorus removal, there can be some impact on PAA decay rates, although there is anecdotal evidence that very low concentrations of aluminum that may bleed through an alum-based coagulant process may improve disinfection by acting as an activator for producing radicals that are powerful disinfectants.

8.5 Biofilm Formation

A light brown/tan growth of biofilm has been observed clinging on contact tank walls when PAA is used (Figure 11.3). Similar to algae on secondary clarifier weirs, this requires periodic cleaning (approximately, once every 2 to 3 weeks). Ongoing studies have identified specific bacterial species including *Acinetobacter calcoaceticus* and *Flavobacterium granuli* present in the biofilms formed in clarifiers and other secondary water handling areas and equipment. These types of bacteria are likely to survive primary sludge

FIGURE 11.3 Biofilm that may form when PAA is used (photo courtesy of Vince Chiu).

bed treatment, produce a biofilm contributing rhamnolipid (which is light brown in color), and are prolific in environments with available nitrogen. Some accumulation of biofilm can be reduced or controlled by elevating disinfectant dose during low-flow periods or by isolating areas of concern and dosing higher levels of PAA.

9.0 REFERENCES

American Society for Testing and Materials (ASTM) D4285-83, *Standard Method for Indicating Oil or Water in Compressed Air.*

Block, P., Morgan, S., Bell, K., & Stewart, S. (2015). Control strategies for PAA wastewater disinfection process with variable effluent quality. *Proceedings of the Annual Water Environment Federation Annual Technical Exhibition and Conference.* Chicago, Illinois, United States. Alexandria, VA: Water Environment Federation.

Henao, L. D., Compagni, R. D., Turolla, A., & Antonelli, M. (2018). Influence of inorganic and organic compounds on the decay of peracetic acid in wastewater disinfection. *Chemical Engineering Journal,* 337, 133–142.

12

Economics of Peracetic Acid

Nicole Stephens, Amy Rushton, Cody Charnas, Andy Burnham, & Shay Coombs

1.0 INTRODUCTION

Economic evaluations are often the most critical component of a disinfection alternative selection process. Because peracetic acid (PAA) systems are most often implemented as retrofits to existing sodium hypochlorite storage and dosing systems, many costs and benefits can overlap between the alternatives or can easily be overlooked. Selecting the right evaluation tool and applying it correctly is essential to ensure an evaluation captures all possible financial impacts for each alternative.

Several key design conditions have a significant impact on the financial viability of implementing PAA at a given facility, including

- the ability to repurpose existing infrastructure (retrofit),
- the PAA chemical supplier procurement process and extent of contracted responsibility,
- facility location relative to chemical production facilities,
- designed usable life on equipment,
- baseline effluent water quality and variations in secondary effluent quality, and
- regulatory and stakeholder approval.

While PAA disinfection offers the potential to reduce disinfection doses, thereby using less chemical on an annual basis, one of the main financial drawbacks of PAA is the chemical unit cost. Depending on the location of a facility, the required dose, and the necessity to use a quenching agent, annual chemical costs associated with PAA disinfection may be higher than that of chlorine disinfection, even if less bulk chemical is being consumed. As PAA becomes more widely used in North America, the production cost may decrease because of competition and more widespread production and vendors, resulting in lower hauling costs.

The following sections provide an overview of available economic evaluation tools and discuss, in detail, the use of a holistic financial model to evaluate PAA disinfection. These sections also offer a specific example of an alternatives analysis between PAA and sodium hypochlorite and important design, procurement, and bidding considerations to be made when evaluating PAA.

2.0 BUSINESS CASE EVALUATION TOOLS

Financial or business case models are widely used to evaluate prospective investment opportunities. These models determine the suitability of investment alternatives by quantifying all applicable costs and benefits. Traditional business case models used to assess investment opportunities include life-cycle cost, cost-benefit, and return on investment analyses.

A life-cycle cost analysis (LCCA) assesses the financial elements of a business case over the lifespan of the proposed investment. A cost-benefit analysis (CBA), shown in Figure 12.1, evaluates the cash flow of a business

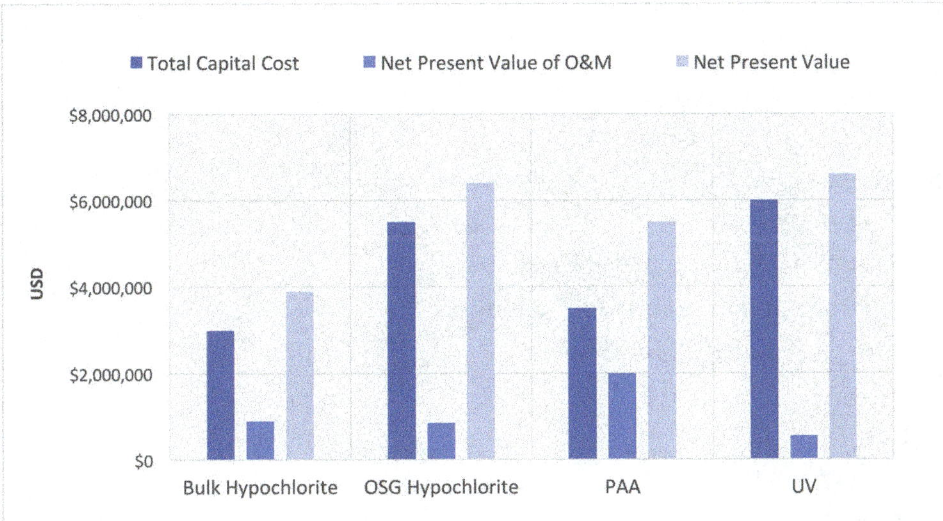

OSG= on-site generation

FIGURE 12.1 Example of a traditional cost-benefit analysis.

investment by expressing each in terms of their net present value. Return on Investment (ROI) analysis is frequently used by organizations to compare the magnitude of costs to the benefits over a specific period.

While traditional business case models account for financial impacts of investments, they tend to ignore social and environmental ramifications of an alternative that can be substantial enough to overturn financial-based decisions. They may also fail to account for uncertainty such as cost overrun and project savings. In 2011, HDR, Inc., with assistance from Columbia University's Graduate School of International and Public Affairs, created a sustainable return on investment (SROI) financial model to holistically evaluate the sustainability of an investment over its entire life cycle.

2.1 Triple Bottom Line

Although the SROI model is similar to LCCAs and CBAs, the SROI employs a triple bottom line methodology. The triple bottom line method monetizes and incorporates all applicable economic, environmental, and social impacts to an investment. In the SROI model, the triple bottom line impacts are divided into the following categories: financial return on investment (FROI), internal sustainable return on investment (ISROI), and external sustainable return on investment (ESROI). Impacts include both costs and/or benefits inherent to a specific investment. When evaluating a disinfection alternative specifically, these categories can include the following subcategories:

FROI

- Capital Costs
- Operations and Maintenance (O&M) Costs

ISROI

- Employee Safety
- Permitting/Regulatory Risks
- Public and Stakeholder Perception
- Project Reliability

ESROI

- Air Quality
- Water Quality
- Conservation
- Transportation
- Odor Formation

2.2 Risk Analysis

The SROI model evaluates sustainability over the lifetime of a project and generates a more realistic evaluation by incorporating uncertainty to the analysis. The risk analysis is combined with a Monte Carlo simulation that uses a random number generator to introduce uncertainty to the model. Model inputs are coupled with a probability distribution and confidence interval, allowing for a range of possible outcomes.

When applicable, inputs are given a range of values including maximum, minimum, and average. Maximum and minimum values are calculated by entering a skew. A skewed distribution is an asymmetrical distribution curve with the average (median) not equal to the mode. The skew represents the degree of uncertainty, so a positive or negative skew determines the likelihood of the value falling to the right or the left of the average. For example, if an alternative is more likely to reduce the O&M costs, then the distribution curve should have a positive skew. A large skew indicates a greater uncertainty in the probable outcome.

3.0 PERACETIC ACID FINANCIAL RETURN ON INVESTMENT

The financial portion of an evaluation quantifies all associated capital and O&M costs of an alternative. Capital costs are one-time expenses, and they should include all design and construction costs as well as any other ancillary costs necessary for the implementation of an alternative. The O&M costs are ongoing expenses necessary to operate and maintain an alternative. Many factors influence the financial burden of a disinfection alternative, such as the ability to retrofit existing facilities, required materials of construction, dose requirements, the state of the industry, and location (i.e., climate impacts on chemical storage, need to generate on-site, chemical delivery costs, etc.).

3.1 Capital Cost

Capital costs should account for all resources and materials necessary for the implementation of each alternative, including engineering, administrative costs, equipment, and construction costs. Escalation of these costs must be applied based on the expected date of construction and time to complete the project. Capital cost inputs may be obtained through proposals from vendors, past contracts, and/or planning documents and engineering estimates. The scope of work assigned to the owner, contractor, equipment vendors, and PAA supplier should be carefully considered, and changes to such scope may result in varying costs between alternatives. For example, major equipment such as bulk storage tanks or chemical metering pumps may be supplied by either the general contractor or the chemical supplier, and costs and availability may vary between the two.

Another capital cost to consider is the labor and materials associated with startup of each alternative because these costs are only realized once over the study period. For PAA disinfection, higher doses may be required during the startup period and more frequent manual sampling may be required for process monitoring. If such costs are included for the PAA disinfection alternative, similar costs must be determined for the startup of all other alternatives that are not the existing disinfection approach.

3.1.1 Equipment Costs

When evaluating PAA disinfection against an existing alternative (i.e., existing chlorine disinfection), any fixed assets that will require replacement during the study period should be considered. This may include buildings, pumps, pipes, valves, and so on that will reach the end of their service life during the study period. However, some equipment may surpass its service life during the study period and an assumption that only a percentage of the assets will be replaced may be appropriate to apply to existing assets costs.

For new disinfection alternatives, major equipment such as tanks or pumps may be procured by the owner or chemical supplier under a contract that is separate from the construction contract. In this case, the equipment and delivery costs are included under the equipment costs category and installation costs are included under the construction costs category.

3.1.2 Construction Costs

Construction costs should consider all materials and equipment not estimated as equipment costs, including pipe and fittings, instruments, pumps, tanks, concrete, structural supports, buildings, electrical, and HVAC. Construction costs can be determined using an opinion of probable construction costs (OPCC) estimate. An OPCC uses a set of assumptions, design recommendations, and known construction constraints to estimate the total construction cost (material and labor) for an alternative. The OPCCs are often classified based on the maturity level of the project deliverables, where a Class 5 estimate reflects the conceptual design phase, and a Class 1 estimate is based on nearly complete or 100% complete design documents.

The accuracy of the estimate is dependent on the level of detail in the design; therefore, design for each alternative should be progressed to similar levels before estimating construction costs. The estimate classification and applied contingency must be consistent between alternatives. In some cases, enough information may be available to develop a Class 3 OPCC for one alternative. However, if only conceptual information is available for the other alternatives, a Class 5 estimate must be used for all to avoid skewing results in favor of the more developed alternative.

The estimate should also delineate which work will be performed by the general contractor and which work will be performed by a subcontractor, such as electrical, instrumentation and controls, and masonry. Labor and equipment rental costs must also be estimated for site civil work (including excavations for pipe trenches) and demolition of existing structures and equipment, where the new disinfection system will be constructed. Contingency is added to the final estimate to account for unforeseen construction costs. The magnitude of the contingency is directly related to the extent of design detail developed for each alternative. Some owners may have a policy on the contingency applied for each OPCC classification.

3.1.3 Engineering Costs

Engineering costs are most often estimated as a percentage of the equipment and construction costs. Engineering costs may include conceptual or preliminary design of alternatives, alternative evaluation and selection, detailed design of the selected alternative, engineering services during construction, and startup and data analysis services. Engineering costs are typically 15 to 35% of the total construction (equipment and construction) costs, depending on the extent of work self-performed by the owner or contracted to an engineering consultant.

Table 12.1 presents a summary of capital cost estimate inputs and calculations. Indirect cost factors shown in the table are typical values and are subject to change based on the project location.

3.2 Operations and Maintenance Costs

Operations and maintenance costs account for all materials, resources, and labor necessary to maintain proper operation of each alternative. The O&M costs can be obtained via equipment vendors, consultants, work orders, and existing utility expenses.

3.2.1 Energy

Energy costs include both power and fuel (typically estimated as electricity and natural gas consumption). They can be calculated based on existing consumption rates and may be escalated for additional future units that are needed for an alternative. The most energy-intensive component of a PAA disinfection system are the chemical metering pumps. Because chemical metering pumps are often used in various other parts of a water resource recovery facility (WRRF), energy usage data from existing pumps can likely be used to estimate that of the new PAA system pumps.

Equipment vendors may also provide energy-consumption estimates. If a specific component of an alternative is known to have significant impacts on energy consumption such as ultraviolet (UV) reactors, vendor estimates for energy consumption must be obtained. The unit cost of energy and the estimated consumption are input as separate line items in the estimate to address variability in energy use and uncertainty in market values of electricity and natural gas.

3.2.2 Chemicals

Chemical costs can be calculated based on increases or decreases to existing dose requirements for a given alternative. Unit costs of chemicals are obtained from vendors, and the change in consumption may be estimated by the vendor or engineer or determined by bench- or pilot-scale studies. Chemical unit cost and estimated consumption rates are entered as independent line items in the estimate to address variability in dose requirements and uncertainty in market values of the chemical.

3.2.3 Labor

All other labor-related O&M costs can be combined into one input. This includes all labor for sampling and analysis, parts replacement, chemical

TABLE 12.1 Capital cost example calculation.

Input	Source/calculation	Unit
Direct costs		
Labor	OPCC[a] estimate	$
Material	OPCC estimate	$
Subcontract	OPCC estimate	$
Equipment	OPCC estimate	$
General conditions	OPCC estimate	$
Indirect costs		
Sales tax	Approx. 4.5% × (material costs)	$
Overhead	Approx. 12% × (direct cost + sales tax)	$
Profit	Approx. 5% × (direct cost + sales tax)	$
Bonds	Approx. 1.3% × (direct cost + sales tax)	$
Insurance	Approx. 1.1% × (direct cost + sales tax)	$
Subtotal A	Direct costs + indirect costs	$
Escalation	Subtotal A × (1+i)^n	$
Date estimate calculated	MM-DD-YY	
Midpoint of construction	MM-DD-YY	
i =	2–5%	%
n =	Date estimate calculated − midpoint of construction/365.25	
Subtotal B	Subtotal A + escalation	$
Engineering	15–30% × (Subtotal B)	$
Legal & administrative	3–7% × (Subtotal B)	$
Subtotal C	Subtotal B + engineering + legal & administrative	$
Contingency	30% × (Subtotal C)	$
TOTAL CAPITAL COST	Subtotal C + contingency	$

[a]OPCC refers to opinion of probable construction costs.

deliveries, process monitoring, system or instrument cleaning, and so on. Work orders can be used to estimate labor for both routine and unscheduled maintenance activities. Assumptions can also be made for new processes based on similar equipment. For example, labor required to operate and maintain a PAA disinfection system may be very similar to that required for an existing chlorine disinfection system because they have many of the same major components such as tanks, pumps, and diffusers. A deduction may be included for PAA labor to capture the reduction in labor that results from having fewer chemical systems, less run time on pumps, and more durable pipe and tank materials. Vendors and consultants can also assist in developing labor costs through previous experience and/or obtaining estimates from clients with similar systems already in service.

Table 12.2 provides a summary of O&M cost-estimate inputs and calculations.

4.0 PERACETIC ACID INTERNAL SUSTAINABLE RETURN ON INVESTMENT

The internal portion of an SROI model integrates all internal, non-monetary impacts of the alternatives. Internal non-monetary impacts are costs and benefits that affect the owner but are not associated with capital or O&M. These ISROI impacts are monetized by determining labor rates and potential hours for responding to these items. Examples of ISROI costs for wastewater treatment alternatives include the following:

- Employee safety—chemical spill response and transportation safety.
- Stakeholder and public acceptance—handling customer questions or complaints.
- Permitting/regulatory risks—site application requirements and permit and regulatory abidance.

4.1 Safety

The safety category should capture safety risks associated with both chemical storage and handling and freight transportation of chemicals. Employee safety costs can be calculated based on average labor rates for a spill response team and an assumption of spill risk based on chemical usage. Employee labor rates and response times are input separately to address variability in labor rates and the probability of a spill. Safety risk related to chemical transport can be monetized by estimating the change in annual chemical hauling miles caused by increasing or decreasing chemical consumption.

TABLE 12.2 Operations and maintenance cost example calculation.

Input	Source/calculation	Unit
Electricity	Consumption × cost	\$/yr
Electricity consumption	Vendor supplied	kWh/yr
Electrical cost	Local rate	\$/kWh
Natural gas	Consumption × cost	\$/yr
Natural gas consumption	Vendor supplied	decatherm/yr
Natural gas cost	Local rate	\$/decatherm
Chemicals	Consumption × cost	\$/yr
Disinfection chemical consumption	Vendor supplied	kg/yr (lb/yr)
Disinfection chemical unit cost	Local rate	\$/kg (\$/lb)
Quenching chemical consumption	Vendor supplied	kg/yr (lb/yr)
Quenching chemical unit cost	Local rate	\$/kg (\$/lb)
Ancillary chemical consumption	Vendor supplied	kg/yr (lb/yr)
Ancillary chemical cost	Local rate	\$/kg (\$/lb)
Operational labor	Labor × cost	\$/yr
Labor required	Vendor/owner estimate	hours/yr
Hourly Cost	Owner labor rates	\$/hr
Maintenance labor	Labor × cost	\$/yr
Labor required	Vendor/owner estimate	hours/yr
Hourly cost	Owner labor rates	\$/hr
TOTAL O&M COST	Electricity + natural gas + chemicals + operations + maintenance	

This value is then multiplied by a cost of accident risk per vehicle mile traveled (\$/VMT). The cost per VMT can be generated from local traffic safety reports and estimates on the value of a statistical life. It is recommended that a third-party study be performed to obtain these values.

4.2 Stakeholder and Public Acceptance

Stakeholder and public acceptance incorporates internal dealings with the governing body (board of directors, city council, etc.), the public, and the regulatory authority. These costs are monetized by estimating the number of hours required to respond to customer calls and the average salary of customer service employees.

4.3 Permitting and Site Application

Permitting and site application inputs account for the costs associated with the effluent discharge permitting process and the assumed cost to prepare a site application during design. Existing system alternatives will not need to include costs for site application and design. Table 12.3 provides a summary of internal cost estimate inputs and calculations.

5.0 PERACETIC ACID EXTERNAL SUSTAINABLE RETURN ON INVESTMENT

The external component focuses on the environmental impact of each alternative and monetizes the detriment or benefit realized by changes in emissions and receiving water quality. This component distinguishes the SROI model from other business case evaluation tools because it allows the user to compare sustainability related costs and benefits, which are typically qualitative, on a financial basis. The most typical impacts considered are greenhouse gas (GHG) and air-contaminant emissions and discharges to receiving waters, as discussed in the following sections. Other potential impacts to consider include conservation efforts, transportation, and odor emissions.

5.1 Greenhouse Gases and Criteria Air Contaminants

Greenhouse gases and criteria air contaminants (CACs) are monetized by the total emissions from production of the chemical, transportation to the site, and operation of the chemical storage and dosing system. Greenhouse gas quantities are assessed based on the amount of carbon dioxide emitted into the atmosphere, and CACs are based on nitrous oxides, volatile organic compounds, particulate matter, and sulfur dioxide emissions. Similar to chemical costs, emission costs and quantities are calculated independently to address variability and in emissions and uncertainty in the estimated cost.

Estimates for emissions produced while making the disinfection chemical or UV reactors should be provided by the chemical supplier. Emissions from transportation of the chemical can be calculated based on the estimated

TABLE 12.3 Internal cost example calculation.

Input	Source	Unit
Safety		
Chemical spill response hours	Change in storage volume and probability of spill	hr/yr
Hourly wage of spill response team	Assumption provided by owner	\$/hr
Increase in freight truck miles	Change in annual chemical consumption	miles/yr
Cost per vehicle mile traveled	Third-party evaluation	\$/VMT
Stakeholder and public acceptance		
Hours for regulatory stakeholder outreach	Assumption provided by owner	hr/yr
Hourly wage of employees	Assumption provided by owner	\$/hr
Public project acceptance		
Hours allocated for public outreach	Assumption provided by owner	hr/yr
Hourly wage of employees	Assumption provided by owner	\$/hr
Permitting		
Facility plan & site application	Assumption provided by owner	\$/yr
Permitting time commitment	Assumption provided by owner	hr/yr
Average hourly wage of employee	Assumption provided by owner	\$/hr

annual chemical consumption, distance between the chemical production facility and the treatment facility, and typical emission rates for the transporting vehicle. Operation emissions should capture both off-gassing from the unit process and any emissions generated from combustion equipment required for the unit process. Operational emissions are not typically applicable for disinfection system evaluations. The unit cost of emissions should be determined for the specific region and is typically obtained from past studies or a third-party evaluation.

5.2 Water Quality

The water quality input is generated by monetizing impacts to downstream users. For example, any additional treatment that a downstream user must institute because of the alternative, including additional chemicals or equipment and increased labor costs, are monetized and included in ESROI input. When comparing PAA disinfection to chlorine-based alternatives, the reduction in sodium, sulfate, and chlorine in the effluent should be considered along with a reduced disinfection byproduct formation potential. Table 12.4 provides a summary of external cost estimate inputs and calculations.

TABLE 12.4 External cost example calculation.

Input	Source	Unit
Water quality		
Change in cost to downstream user	Estimated change in labor, materials, or equipment	\$/yr
GHGs and CACs		
Carbon dioxide produced	Production + transportation + operation emissions	ton/yr
Carbon dioxide impact cost	Third-party evaluation	\$/ton
Sulfur dioxide produced	Production + transportation + operation emissions	ton/yr
Sulfur dioxide impact cost	Third-party evaluation	\$/ton
Nitrous oxides produced	Production + transportation + operation emissions	ton/yr
Nitrous oxides impact cost	Third-party evaluation	\$/ton
Particulate matter produced	Production + transportation + operation emissions	ton/yr
Particulate matter impact cost	Third-party evaluation	\$/ton
Volatile organic carbon produced	Production + transportation + operation emissions	ton/yr
Volatile organic carbon impact cost	Third-party evaluation	\$/ton

Notes. GHG = Greenhouse gas. CAC = Critical ambient contaminants.

6.0 ECONOMIC COMPARISON WITH OTHER DISINFECTANTS

The following sections provide cost and benefit considerations for disinfection alternatives relative to PAA.

6.1 Chlorine Gas with Dechlorination

Compared to PAA, chlorine gas poses a significantly higher safety risk to operations staff and residents living near the facility. Chlorine gas can cause fatalities in the event of a leak and requires a risk management plan (RMP) and appropriate monitoring equipment to detect leaks. Although PAA is a known skin and eye irritant, it eliminates many safety risks and costs associated with maintaining an RMP. Chlorine gas is stored in pressurized cylinders that must be installed in a controlled climate building with appropriate explosion prevention, which adds significant capital cost compared to a PAA system that can be installed outdoors in most climates.

Capital costs for a chlorine gas storage and dosing system are typically higher than those for PAA because the storage and dosing system for chlorine is more complex, and the contact basins are typically larger because of longer required contact times. Chlorine residuals must be quenched before discharge and, therefore, capital costs must be added for an additional chemical storage and dosing system and associated O&M costs must be included.

Chlorine will react with organics in the water to produce chlorinated disinfection byproducts (DBPs). Although DBPs are not currently regulated for effluent discharge, additional treatment costs should be included in the ESROI portion of an evaluation to account for the resulting effluent toxicity or the impact to downstream users.

6.2 Sodium Hypochlorite with Dechlorination

Sodium hypochlorite (SHC) may pose a higher safety risk to O&M staff than PAA. Higher chemical feed rates require larger bulk storage tanks, resulting in increased spill and exposure risk. Response time and costs for a SHC spill will be higher than PAA because of larger storage volume requirements and the persistence of chlorine in the environment if containment systems fail.

Sodium hypochlorite is less stable than PAA and continuously off-gasses in storage. Proper ventilation on tanks and valves and the monitoring system must be included in capital cost estimates. The higher instability also results in more frequent chemical deliveries and a potential to underdose if bulk chemical is stored too long and the active SHC concentration degrades below assumed strengths. More frequent deliveries increase the safety risk to the surrounding community, and the increases in GHG emissions resulting from off-gassing should be considered.

The SHC must be stored in climate-controlled buildings to prevent high temperatures from exacerbating degradation and low temperatures from freezing the chemical. Capital costs for SHC storage and dosing systems are typically similar to that of PAA. Although an SHC system will be much larger to account for more bulk chemical consumption, storage tanks are typically fiberglass, and pipe and fittings are either chlorinated polyvinyl chloride (CPVC) or a thermoplastic, which are typically cheaper than the stainless steel required for PAA systems.

On-site generation of SHC results in even higher capital costs because of additional equipment costs, larger bulk tanks, and larger feed pumps attributable to the low SHC concentrations produced. However, on-site generation systems can have lower O&M costs than bulk SHC systems. At many facilities, chloramination (which uses ammonia and sodium hypochlorite to generate chloramines) is used in lieu of free-chlorine disinfection because it reduces the chlorinated DBP formation potential. In this case, the costs for capital, fixed-asset replacement, and O&M of an ammonia facility must also be included.

Although chlorine-based disinfection systems typically require more chemicals and higher feed rates than PAA, SHC systems can have lower annual O&M costs. This is primarily an artifact of higher chemical costs for PAA because of a limited number of production facilities in the United States, and it varies by location. For example, the distance that PAA must travel to a facility can have a significate impact on its unit cost. However, with PAA gaining attention and demand increasing in the market, it is expected that the number of PAA suppliers will increase and prices will start to decline.

Like chlorine gas, a quenching is also required for sodium hypochlorite disinfection and the water quality burden of DBP formation must be considered.

6.3 Ultraviolet Disinfection

Ultraviolet disinfection can be a safer alternative to PAA because it eliminates the storage and handling requirements that accompany the use of chemicals. The risk of a chemical spill or gas exposure to surrounding areas is eliminated with UV systems because pathogens are inactivated by UV exposure rather than being chemically oxidized. The reduction and/or elimination of these risks may result in lower safety costs for UV systems when compared to PAA. However, one important safety and environmental burden that should be monetized is the risk of mercury contamination caused by lamp breaks.

The treatment reliability of UV disinfection is more dependent on upstream conditions than PAA, with turbidity having a large impact on treatment efficiency. Turbidity spikes often require high UV doses to inactivate harmful microbes, resulting in higher energy use or permit violations if control systems do not respond appropriately. In some cases, it may be most cost-effective or necessary to include filters upstream of UV systems for optimal disinfection.

Because the size of a UV facility may be drastically different when sized for peak flow compared to maximum month flow, a facility might find it most feasible to design their UV system for only maximum month conditions and supplement with chemical (PAA) disinfection during peak flow events. This alternative may provide an optimized balance between the PAA and UV system sizes and consumption of consumables. Constructing a new chemical disinfection facility in addition to a new UV disinfection facility would be a significant capital investment and would not be justifiable against a single UV facility sized for peak flow. Therefore, this configuration is most appropriate for a facility that can retrofit existing chemical storage and contact infrastructure for PAA.

Retrofitting a UV system into existing chemical contact tanks typically involves structural modifications to the existing channels to fit the UV modules or reconfigure the flow path. Peracetic acid systems can be retrofit at an existing facility with no modifications to the existing contact tanks, resulting in comparatively higher capital costs for a UV retrofit.

The O&M cost estimates for UV systems are fairly simple and straightforward because electricity is the major consumable input. Operations and maintenance cost estimates can be easily obtained from equipment vendors and should include electricity consumption, lamp cleaning (including chemical consumption if acid cleaning is used), lamp replacement, and ballast replacement.

Unlike chlorine-based systems, DBP formation is not a concern, and UV disinfection will typically have the lowest ESROI burden because it does not require addition of any compounds to the effluent.

6.4 Ozone

Ozone disinfection typically requires the most significant capital investment. Ozone disinfection facilities can make use of existing buildings and contactors; however, a substantial amount of additional infrastructure is required to make contact basins airtight and ozone-containing facilities safe. Ozone systems require far more equipment compared to other disinfection alternatives and, at a minimum, must include liquid oxygen storage and vaporization or high-purity oxygen generation, ozone generators and power supply units, cooling water system, injection system, contact basins, off-gas destruction system, ambient monitoring systems, and a quenching system. If a new facility is being constructed, smaller contact basins would be required for ozone compared to chlorine and PAA; however, they need to be capped, well baffled, and maintained under vacuum.

Ozone O&M costs can also be significant because the generators require substantial power inputs, several large pumping systems are required, and a quenching chemical is needed to prevent discharging ozone residuals and potential human exposure through off-gassing. There is also a significant health and safety risk because ozone is highly toxic to humans at very low concentrations. However, some financial credit may be assigned to ozone in an evaluation for its ability to improve water quality beyond pathogen removal. While chlorinated DBP formation is not a concern, brominated DBP formation potential must be considered.

7.0 METRO DISTRICT SROI ANALYSIS OF CHLORAMINES AND PAA

This section describes a disinfection alternative evaluation performed by the Metro Wastewater Reclamation District (Metro District) for its 833-mL/d (220-mgd) Robert W. Hite Treatment Facility. In 2016, the Metro District prepared a SROI model to determine the feasibility of PAA as a wastewater disinfection alternative to continued use of their existing sodium hypochlorite (chloramination) approach. The Metro District uses 12.5% SHC and 15% aqua ammonia for chloramine disinfection. Sodium bisulfite (SBS) is added to neutralize the residual chlorine before discharge.

The Metro District also included UV disinfection and a combination of UV and PAA disinfection into the evaluation. The UV and PAA alternative assumed UV would treat the majority of the effluent and PAA would only treat a sidestream, particularly during peak wet weather flows. The SROI model allowed the Metro District to monetize all relevant costs and benefits to accurately compare project alternatives.

It was determined that PAA equipment would be leased through the PAA chemical supplier rather than purchased by the contractor or owner. Therefore, PAA equipment costs were included under chemical costs rather than equipment to reflect the lease agreement. The capital costs for the existing disinfection system were determined by assuming that existing fixed assets for SHC, SBS, and ammonia systems would reach the end of their service life before the end of the study period. The total cost was then reduced by 50% under the assumption that only half of those assets would be replaced during the study period, consistent with Metro District's fixed-asset budgeting process. Capital costs for the UV and UV/PAA alternative were derived from the Metro District's latest facility plan. All costs were escalated to reflect current market value.

Energy cost estimates used the current bulk rate at the Metro District. Energy requirements for the PAA alternative were based on vendor-provided information. Utility billing was used for the existing disinfection system, and UV energy demands were developed with the assistance of a consulting firm.

As mentioned previously, chemical costs in the FROI analyses for the PAA alternative included all equipment costs, and the unit cost of the PAA was provided by vendors. Chemical costs for the existing chloramine disinfection system were based on dosing requirements and current unit costs. There were no chemical costs associated with the UV alternative.

The Metro District worked closely with the O&M and environmental services staff to develop O&M costs for each alternative. The O&M costs included labor for operations, routine maintenance and sampling, and chemical handling. Labor requirements for O&M, administration, and laboratory activities were developed using historical work orders. Average labor rates were calculated for each department and input to the model for the analysis.

The ISROI analysis included monetized inputs for safety, public acceptance, and permitting. Safety inputs were developed using response time and cleanup activities required during a spill scenario for each alternative. Spill response for PAA was compared against those of SHC, SBC, and ammonia. Public acceptance inputs were similar for all alternatives. However, inputs for the existing disinfection system were slightly higher to account for the risks associated with a SHC permit violation and the associated notifications. Site application costs were included for all alternatives, except the existing disinfection system.

7.1 Results

Figure 12.2 shows the results of the financial portion of the alternatives analysis. This includes the total net present value (NPV) of capital and O&M costs only. When strictly financial burdens are considered, PAA disinfection was the most desirable alternative, followed closely by SHC disinfection. In addition, all the alternatives considered had a negative total NPV, meaning there is no payback. However, when the ISROI and ESROI components of the evaluation are added in, as shown in Figure 12.3, SHC disinfection becomes the least desirable alternative and PAA shows a positive return on its investment. The probability of not exceeding reflects the uncertainty of the estimated NPVs based on the risk analysis.

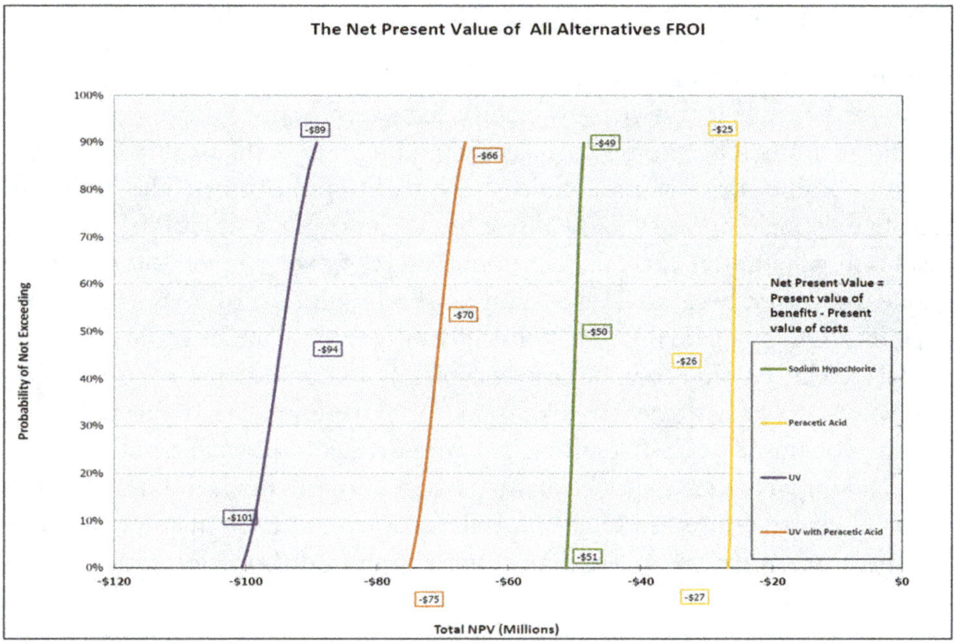

FIGURE 12.2 Metro District net present value (NPV) of all alternatives financial return on investment (FROI).

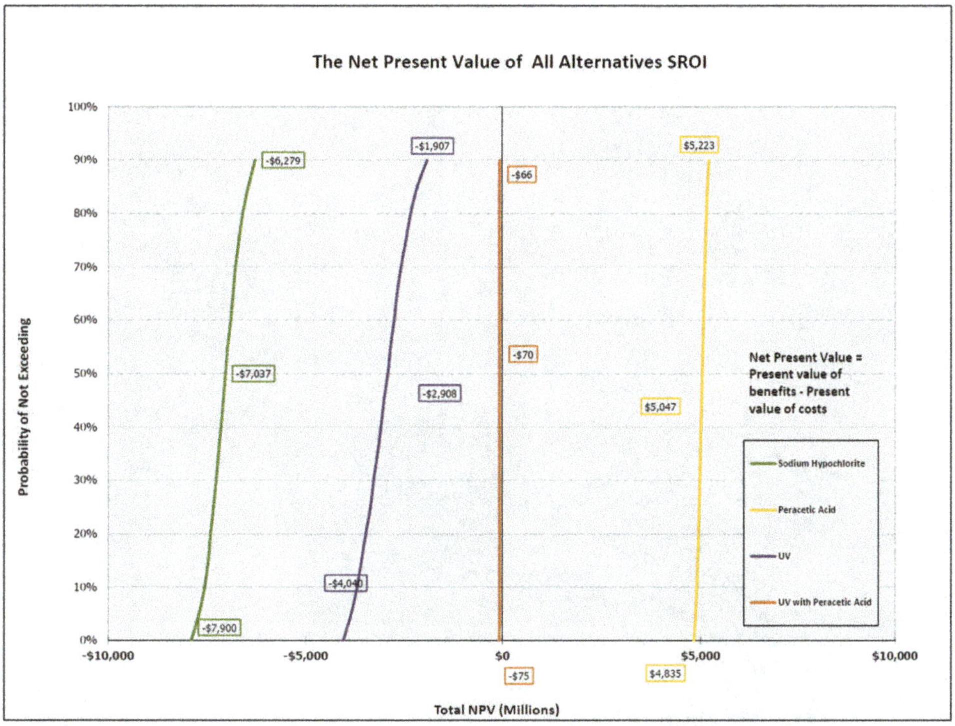

FIGURE 12.3 Metro District NPV of all alternatives SROI.

7.2 Summary

Peracetic acid continues to gain popularity as a municipal wastewater disinfectant because it can cost-efficiently remove pathogens to meet effluent discharge limits. The capital costs associated with PAA disinfection are typically favorable for both greenfield and retrofit applications. The ability to retrofit existing chlorine chemical storage, dosing, and contact infrastructure results in substantial capital investment savings over conversion to UV or ozone. However, because of the relatively high unit cost of PAA, the annual expenditure may make PAA economically unfavorable under traditional cost analyses. Many of the considerations typically evaluated under the "qualitative" portion of an alternatives analysis can have financial impacts and can actually be monetized. Therefore, it is important to consider not only the financial impacts of a disinfection alternative, but the potential internal and external returns on investment as well. In doing so, PAA disinfection may become even more favorable than more common disinfection alternatives.

13

Peracetic Acid Implementation Case Studies

Kati Bell, Brian Hilts, Sarah Stewart, Michael Watts, Greg Eyerly, Reynold Dale Richwine, Phyllis Frank, & Walter Collins

1.0 INTRODUCTION

This manual aims to advance the knowledge of peracetic acid (PAA) disinfection and support the development of guidance for design, implementation, and operations. Although only a few dozen facilities globally use PAA disinfection at the time of this publication, it is important to gather information from these projects to identify technical advances that have been made and lessons that have been learned. This chapter compiles existing information on six case studies relating to the implementation of PAA disinfection for wastewater effluent that exemplify the information presented throughout this document. These case studies are intended to serve as tools for utilities, regulators, engineers, and operators to provide site-specific data, management perspective, and operational experience.

Additionally, the case studies highlight how PAA can be used as a cost-effective solution that addresses many challenges related to meeting bacterial compliance while also minimizing risks associated with traditional gas chlorine-based disinfection and protecting both human and environmental health. Table 13.1 provides a summary of the five case studies presented herein and how these each inform various aspects of the implementation of PAA disinfection. The case studies themselves are provided in the following sections of this chapter.

2.0 NORTHWEST LANGLEY WASTEWATER TREATMENT PLANT

2.1 Background

The Northwest Langley Wastewater Treatment Plant (WWTP) is a trickling filter facility that has an average daily flow of 12 million liters per day (MLD), with peak flows reaching approximately 18 MLD; Figure 13.1 provides a process flow diagram. The treated effluent is discharged into the Fraser River and is regulated under the Fisheries Act, which prohibits unauthorized deterioration, disruption, and destruction of fish habitat. The regulations specifically address limits for final effluent quality, including total residual chlorine (TRC). In efforts to address the risks associated with noncompliance of TRC, Metro Vancouver investigated the use of PAA at the Northwest Langley WWTP starting in 2012.

2.2 Pilot Testing

Temporary PAA chemical storage and feed facilities (Figure 13.2) were constructed to support pilot testing, which was conducted to determine the

TABLE 13.1 Summary of PAA implementation case studies (Table created by Kati Bell).

Owner facility name location	Average flow/ peak flow	Drivers	Startup date
Metro Vancouver Northwest Langley WWTP Vancouver, BC	3/5 mgd 12/18 mL/d	Risk of violating total residual chlorine limit and toxicity testing noncompliance	2012
Tullahoma Utilities Board Tullahoma WWTP Tullahoma, TN	5/30 mgd 19/113 mL/d (SBR decant)	Low capital cost alternative to gas chlorine and sulfur dioxide	2014
Gulf Coast Authority Washburn Tunnel Pasadena, TX	16/43 mgd 61/163 mL/d	New system to meet Enterococcus limits on a footprint constrained site	2016
City of Memphis M.C. Stiles WWTP Memphis, TN	135/250 mgd 511/946 mL/d	New system to meet *E. coli* limits without halogenated disinfection byproducts	2018
Little Rock Water Reclamation Authority Adams Field WWTP, Little Rock, AR	36/72 mgd 136/273 mL/d	Supplemental feed for underperforming UV disinfection system that experiences low water quality at high flows	2016

required doses for permit compliance and to evaluate any possible toxicity. Full-scale piloting was conducted in 2013 and 2014 during the disinfection season (April 1 through October 31). Results of the Stage 1 testing indicated an initial dose of 2 mg/L of PAA (as active ingredient) was sufficient to reduce effluent fecal coliform concentrations to less than 700 MPN/100 mL. To advance the understanding of PAA disinfection at the Northwest Langley WWTP and build on the initial testing, additional testing (Stage 2) was conducted in 2014.

In addition to the regulatory considerations with respect to bacteria compliance and TRC, there are also other risk management considerations that may not have a direct measurable cost. Some of these considerations include:

- Improved process control compared to chlorine because of the issues associated with nitrite formation as the result of partial nitrification;

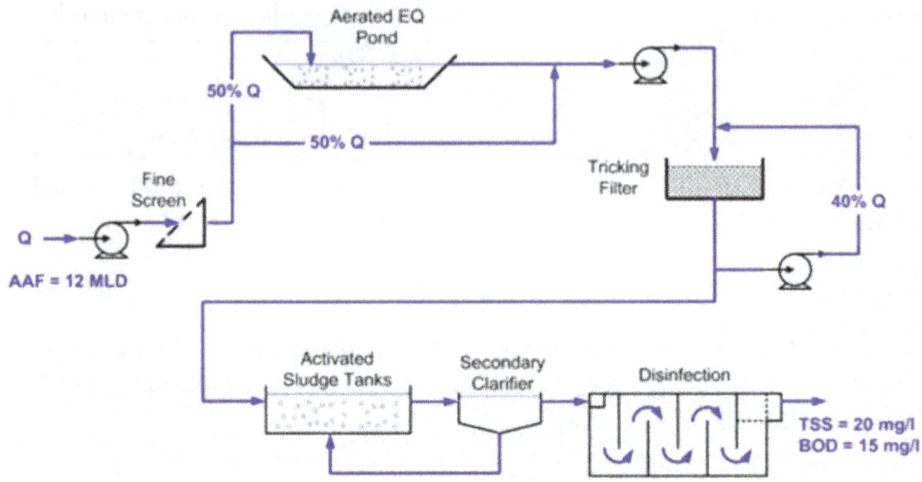

FIGURE 13.1 Process flow diagram of the Northwest Langley Wastewater Treatment Plant (courtesy of Vince Chiu).

FIGURE 13.2 Temporary PAA chemical storage and feed facilities to support piloting at the Northwest Langley Wastewater Treatment Plant (photo courtesy of Vince Chiu).

- The ability of PAA to address emerging contaminants of emerging concern such as endocrine disrupting compounds; the improvement of effluent color; and,
- Elimination of the need to dechlorinate, which is an important risk management consideration for compliance and improved effluent toxicity characteristics.

Thus, in addition to developing data to support development of a recommended design dose, a substantial dataset of aquatic toxicity data was collected. This dataset has been compiled and reviewed as part of understanding PAA toxicity with respect to its implementation as a disinfectant at the Northwest Langley WWTP.

2.3 Improved Effluent Toxicity Characteristics

At the time of publication of this manual, there are yet only a few dozen water resource recovery facilities (WRRFs) in North America using this technology for disinfection, although both the Pest Management Regulation Agency (PMRA) of Canada and the U.S. Environmental Protection Agency (U.S. EPA) have approved PAA for use as a disinfectant to treat wastewater. Notably, St. Augustine and Flagler Beach, Florida, have implemented PAA to address formation of dichlorobromomethane limits in their National Pollutant Discharge Elimination system (NPDES) permit. As described in the City of Memphis case study in this chapter, PAA was selected because it could provide bacterial compliance without formation of dioxin and other chlorinated disinfection byproducts (DBPs). As indicated by implementation of the case study projects, PAA is an attractive alternative disinfection technology because it does not produce the regulated DBPs of concern that are associated with chlorine disinfection.

The research that has been published to date with respect to toxicity of wastewater and surface water disinfected with PAA has not been shown to result in notably toxic or mutagenic byproducts (Baldry and Fraser, 1988; Kitis, 2003; Monarca et al., 2002; Monarca et al., 2001). Thus, from a regulated or unregulated DBP perspective, PAA is more favorable than chlorine disinfection. And, while PAA does not produce regulated DBPs, Metro Vancouver wanted to develop data to address any questions related to related toxicity.

2.4 Toxicity Testing at the Northwest Langley Wastewater Treatment Plant

Before any testing at the Northwest Langley WWTP, Metro Vancouver established toxicity profiles of PAA for rainbow trout to inform the development of the test plan that included two phases of work. It is of note that during Stage 1 of the project, bench-scale laboratory tests were conducted to determine the 96-hour lethal concentration at which 50% mortality occurs (LC_{50}) threshold concentration of PAA for two vendors.

In summary, the laboratory established a series of 10 dilutions to establish an approximate range for the toxicity threshold; then, full 96-hour LC_{50} tests (rainbow trout) were conducted on five doses of PAA to establish the maximum effluent PAA residual that would not be toxic. The results of the testing showed that the 96-hour LC_{50} could be estimated to be 4.2 mg/L (as PAA). It is important to note that these results do not mean that PAA doses greater than 4 mg/L are toxic. When PAA is added to effluent, its reaction with constituents in the water, combined with its auto-decay, results in reduced concentration below LC50 levels.

Considering the data that were collected during Stage 2 demonstrating that the average PAA demand of effluent at the Northwest Langley WWTP is 1.7 mg/L, theoretically, a PAA dose of 5.9 mg/L would result in PAA residual of 4.2 mg/L. Also, based on the data collected during Stage 2, it was demonstrated that the expected maximum dose of PAA that would need to be applied is approximately 3.0 mg/L, which would result in a residual of only 1.3 mg/L. In either case, both the maximum dose and maximum residual that would be expected are much less that the threshold value for the 96-hour LC_{50}.

During Stage 2 pilot testing, an additional 40 whole effluent toxicity (WET) tests were performed on PAA-treated effluent using rainbow trout. The tests were conducted with pH stabilization in conjunction with the reference method EPS 1/RM/13, *Biological test method: Reference method for determining acute lethality of effluents to rainbow trout* (Environment Canada, 2008). This method was used because, in wastewater effluent samples, the carbon dioxide (CO_2) content can be elevated as a result of high biological activity.

Aeration of these samples may cause the pH to rise because of a loss of CO_2, and the change in pH can alter the toxicity profile associated with ammonia present in the sample. Thus, the purpose of pH stabilization is to replace the CO_2 lost as a result of aeration for maintaining the pH throughout the test. To use this add-on procedure, the effluent sample must meet three conditions: (a) total ammonia must be measured on all samples submitted for testing, (b) the effluent must have failed an acute lethality test on a previously collected sample, and (c) pH stabilization may only be used when the unionized ammonia in the effluent does not equal or exceed 1.25 mg/L at 15 °C (59 °F) or as determined by a formula outlined in EPS 1 /RM/50, *Procedure for pH stabilization during the testing of acute lethality of wastewater effluent to rainbow trout* (Environment Canada, 2008).

Using these procedures, all WET tests passed the permit criteria; there were a few fish mortalities, however, these mortalities were not statistically significant and were likely related to the test conditions as opposed to any toxicity related to PAA. To further demonstrate that the reported fish mortalities were unrelated to PAA dose, the fish mortalities that were reported were plotted against the corresponding PAA dose to determine if there were any correlation. Results plotted in Figure 13.3 demonstrate the lack of correlation between PAA dose and fish mortality.

2.5 Economic Evaluation and Implementation

Following the successful evaluation of PAA toxicity, life-cycle costs were estimated based on the total construction capital and annual operations cost for PAA compared to chlorination/dichlorination. Table 13.2 presents a summary of the construction costs, annual operations costs, and 20-year life-cycle cost for chlorination/dechlorination using bulk liquid sodium hypochlorite and sodium bisulfite compared with disinfection using PAA.

The cost estimates demonstrated that PAA disinfection has economic benefits with respect to both initial construction costs as well as the life cycle of the project. Thus, in addition to allowing Metro Vancouver to manage risks associated with discharges of TRC, which are strictly limited into the Fraser River, the Northwest Langley WWTP converted temporary

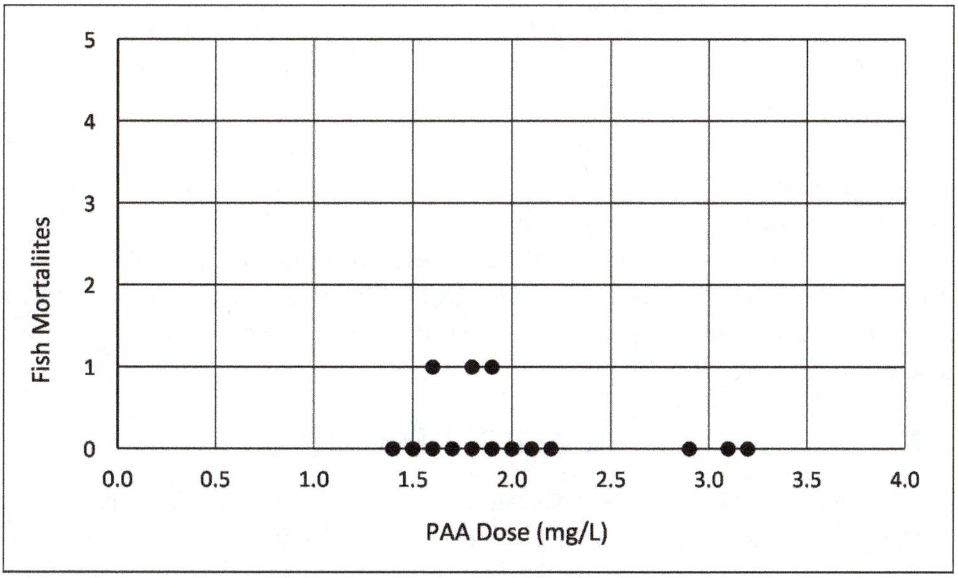

FIGURE 13.3 Correlation between PAA dose and fish mortality (adapted from CDM Smith, 2015).

TABLE 13.2 Estimated annual O&M costs for disinfection alternatives in 2015 $ CSD (adapted from CDM Smith, 2015).

Process alternative	Total construction cost ($ CAD)	Annual operations ($ CAD)	20-year life cycle ($ CAD) *
Chlorination/ dechlorination	2 000 000	68 000	3 500 000
PAA	1 260 000	82 000	3 100 000

*Interest rate = 5%, inflation rate = 10%, 20-year life cycle

facilities into a permanent installation and has been using PAA to meet its disinfection compliance without issues of toxicity or concerns related to TRC noncompliance risks.

3.0 TULLAHOMA UTILITIES BOARD WATER RESOURCE RECOVERY FACILITY

3.1 Background

Tullahoma's water resource recovery facility (WRRF) was installed in 1985 and, at the time, it was the first sequencing batch reactor (SBR) treatment system of its kind in the United States and the largest in the world. Also at the time, it was supported by the U.S. Environmental Protection Agency (U.S. EPA) as an "innovative alternative" to existing systems, and 85% of its construction was funded by grant money (Lapczynski, 2016).

The WRRF, operates and is regulated under a NPDES permit that allows discharges into Rock Creek. The facility, designed for flows of 19 mL/d (5 mgd), was updated in 1995. Since construction of the facility, the WRRF has used chlorine for disinfection and, because of residual limits in the NPDES permit to protect Rock Creek, a low-flow stream with a sensitive ecosystem, the NPDES permit requires dechlorination before discharge (Young, 2016).

3.2 Drivers for Peracetic Acid Disinfection

The WRRF has always experienced maintenance and operational challenges with the gas chlorination and dichlorination equipment, despite using several different equipment systems and assistance from multiple contractors. These issues have been attributed to the nature of operating an SBR, which processes wastewater flows in batches, requiring the disinfection system to turn on and off approximately 12 times per day and adjusting to flows that range from zero to 114 mL/d (30 mgd) in a short period of time.

In addition to operational challenges, the WRRF staff and Tullahoma Utilities Board (TUB) management were required to have a risk management plan (RMP) with significant administrative requirements. Thus, in addition to the operational and administrative complexities of gas chlorine, in March of 2015 an accidental release of chlorine gas at a major WRRF in Chattanooga, Tennessee, further highlighted TUB's concern regarding the potential safety and security risks associated with gas chlorine and sulfur dioxide. The Tullahoma Utilities Board identified PAA as a potential disinfection alternative so as to avoid the high cost of equipment replacement and many manhours for maintenance, along with the safety issues and RMP requirements for handling hazardous (Young, 2016).

3.3 Study and Evaluation Process

The TUB worked closely with the Tennessee Department of Environment and Conservation (TDEC) to put a plan into place to pilot test PAA. The PAA chemical storage and feed system were provided as a pre-engineered system from one of the major chemical suppliers. The system included a chemical storage tote assembled on spill-containment tanks that was connected to a pump cabinet with two peristaltic metering pumps, pressure-relief valves, interconnecting piping, and pump controls. A signal from the WRRF's Parshall flume was used for flow control, with the chemical injection point being immediately downstream of the Parshall flume, ahead of the existing chlorine contact tank.

The pilot was run using various operating conditions, and dosing rates of 0.6 to 1.5 mg/L were shown to provide bacterial compliance, which is to provide a geometric mean *Escherichia coli* concentration of <126 CFU/100 mL as a monthly geometric mean. Additional toxicity testing was conducted to address TDEC concerns related to chronic toxicity of PAA residuals (Young, 2016).

3.4 Permitting Peracetic Acid for a Low-Flow Stream Discharge

In response to TDEC concerns about the receiving stream, chronic WET tests were conducted with *Pimephales promelas* (fathead minnow) and *Ceriodaphnia dubia* (waterflea). The WET results were used to predict the concentrations that interfere with normal growth, development, and reproductive potential of these aquatic organisms. During the chronic test, the organisms were continuously exposed to the effluent at various concentrations (dilutions) over a 7-day period, during which responses such as growth,

FIGURE 13.4 Peracetic acid chemical storage and feed system at the Tullahoma Wastewater Treatment Plant (courtesy of PeroxyChem).

reproduction, and survival were measured. The tests encompass the entire life cycle or most sensitive life stage of the organisms, and effects of chronic toxicity tests are estimated based on the calculation of the concentration at which there is inhibition (in growth or reproduction) of 25% of the organisms in the test (IC_{25}). Results of this test showed the IC_{25} for *C. dubia* to be 0.38 mg/L and for *P. promelas* to be 0.75 mg/L.

The results of pilot testing at the facility also indicated that the hydrogen peroxide component of the product was consistently present at the outfall. Because a majority of the outfall samples for PAA were nondetectable, TDEC asserted that it did not believe that hydrogen peroxide analysis was affected through PAA interference. There is limited available hydrogen peroxide toxicity data, but information available for *Daphnia magna* (0.340 mg/L lowest observed effect level growth) indicated that the levels present in the effluent could be a concern related to aquatic life toxicity.

3.5 Implementation and Peracetic Acid Disinfection System Performance

After more than a year's study into the new wastewater disinfection process, TUB received state approval to replace the chlorine used for wastewater effluent disinfection with PAA. The permit requirements, in addition to *E. coli* limits, also established a reporting protocol for PAA and hydrogen peroxide residual in the discharge and in the receiving water. The TDEC used the IC_{25} results of the most sensitive organism (*C. dubia*) and calculated the allowable instream waste concentration according to the U.S. EPA registration label of the PAA product proposed for the project.

With the data showing that the IC_{25} toxic concentration for *C. dubia* was 0.38 mg/L, the proposed permit limitation for PAA, based on the stream dilution factor of 1.09, was permitted at 0.41 mg/L. Because of the lack of previous experience with permitting PAA, as part of the permit modification, the facility was also required to monitor for hydrogen peroxide as well as conduct periodic instream bioassessments. With that approval, TUB's WRRF was the first in Tennessee to fully implement the use of PAA.

Upon permit renewal, the data collected on instream PAA and hydrogen peroxide were used to establish an enforceable requirement in the most recent NPDES permit, effective December 2018. While instream values for PAA and hydrogen peroxide have been set at 0.38 mg/L and 0.63 mg/L, respectively, these concentrations will be alerts at the outfall instead of being enforceable values. If the sampled result exceeds the alert value, the facility shall initiate corrective action measures, such as reducing the PAA dose to reduce the instream concentration below the alert value. The corrective action measures shall be detailed in a monthly report that is attached to the monthly discharge monitoring report.

3.6 Lessons Learned and Big Benefits to Staff

Conversion of the Tullahoma WRRF has been a successful endeavor that allowed the TUB to eliminate hazards associated with the use of gas chlorine and sulfur dioxide. Both of those dangerous gases have been removed from the facility in favor of liquid PAA, which reacts quickly and dissipates rapidly while providing disinfection compliance. Because it does not persist in the environment or affect effluent toxicity, it does not need to be quenched which is a requirement for chlorination (dichlorination).

Operations staff have reported the system to be simple because "it's literally just a little pump and a tank, where before we had masses and masses of equipment. And we were working on it all the time" (Lapczynski, 2016). Additionally, while, in 2016, the product cost 91 cents per pound, which was higher than the cost of chlorine gas on a per pound basis, the savings incurred by retiring the outdated equipment—both in maintenance and replacement costs—has offset the expense of PAA.

Removing chlorine from the treatment facility also represented an additional benefit for TUB in that moving away from the use of hazardous gases also removed the facility from U.S. EPA's Risk Management Program for toxic or flammable substances—a program which requires regular reporting and annual training with the city fire department.

4.0 WASHBURN TUNNEL WASTEWATER TREATMENT FACILITY

The Washburn Tunnel Wastewater Treatment Facility (Washburn Tunnel), located in Pasadena, Texas, is a regional WRRF owned and operated by the Gulf Coast Authority (GCA). Influent wastewater is comprised of approximately 90% industrial wastewater from refineries, chemical production facilities, and chemical and petrochemical storage facilities, with a minor amount of municipal wastewater influent.

The treatment process is a two-step, single-sludge activated sludge system. The small fraction of municipal wastewater is combined with other high-strength influent streams and treated in a pure oxygen activated sludge process; the other low-strength and high solids wastewater streams are sent to a primary clarifier. Flow from the primary clarifier and the pure oxygen process are combined and split into two air activated sludge treatment trains followed by secondary clarification and discharged into the Houston Ship Channel.

The discharge permit has requirements for effluent enterococci of 168 CFU/100 mL as a 30-day geometric mean, and a maximum of 500 CFU/100 mL based on a daily grab sample. Historically, Washburn Tunnel was able to meet these limits with pretreatment; however, beginning in 2012, the facility experienced elevated enterococci concentrations in the effluent, which necessitated the need to add disinfection to the treatment process.

4.1 Disinfection Evaluation and Testing

The GCA conducted an evaluation of disinfection technologies that could be implemented to achieve bacteria compliance. Eight disinfection technologies were identified for the evaluation: chlorination with dechlorination, chloramination with dechlorination, chlorine dioxide, ozone, ultraviolet (UV) disinfection, PAA, ferrate, and advanced oxidation processes. After an initial desktop screening, GCA conducted bench- and full-scale testing of PAA as well as chloramination/dechlorination.

Based on bench test results, operations staff constructed two 25-gpm pilot reactors (Figure 13.5) to conduct side-by-side testing to determine the design of the disinfectant dose and contact time required to meet permit compliance. Table 13.3 summarizes the contact times and dose ranges for chlorination/dechlorination and PAA disinfection that were tested.

The lower contact time required for PAA would result in a smaller contact tank, reducing overall capital cost and minimizing site impacts for PAA implementation. Although PAA has been implemented in many facilities across the country without quenching, it was included in this project to address the complexities associated with modification of the industrial Texas Pollutant Discharge Elimination System (TPDES) to remove the total residual chlorine test, which has interferences when PAA is present in the sample.

4.2 Peracetic Acid System Design

The new disinfection system was constructed on a severely footprint-limited site. The engineering team collaborated with GCA during design to locate the system, develop layout and configuration of the contact tank, and identify elements of design. The final project included the following:

- an asymmetrical contact tank, with two separate contact channels that allowed the tank to maintain the required 7.5 minutes of chemical contact time at varying facility flows;

FIGURE 13.5 Side-by-side PAA and chlorination pilot reactors (photo courtesy of Kati Bell).

TABLE 13.3 Peracetic acid and chlorination doses and contact times pilot results (adapted from CDM Smith, 2013).

Disinfectant	Contact time (minutes)	Dose range (mg/L)
PAA	7.5	3.5–8
Chlorination	15	15–100

- a PAA feed and mixing system, including two bulk storage tanks and two chemical feed skids, each dedicated to one channel;
- temporary sodium bisulfite (SBS) storage and feed facilities to quench PAA until a long-term solution was identified; and
- a Parshall flume with an ultrasonic level transmitter to provide accurate flow measurement while introducing minimal additional head.

4.3 Peracetic Acid System Procurement

After interviewing PAA suppliers and evaluating various options for procuring PAA and related equipment systems, GCA opted to include an equipment lease contract as part of a turnkey chemical supply contract for PAA to be installed by GCA or its third-party contractor. Chemical supply requirements included all chemical handling equipment and related appurtenances, delivery response commitments, and local backup supply provisions. The GCA used a competitive sealed-bid proposal process to select the PAA provider, which allowed each bidder to be evaluated based on price and performance.

4.4 Peracetic Acid System Construction

The GCA delivered the project using the construction-management-at-risk method to reduce the overall project schedule and obtain a guaranteed maximum price for construction. Constructability reviews and coordination for maintaining facility operations during construction were over a year in planning. Construction commenced in early 2015 and the system was started up in January 2016. Figure 13.6 shows the completed chemical storage and feed facilities.

4.5 PAA System Startup and Optimization

Once the system passed functional testing requirements, process startup testing was performed for 6 months to optimize the PAA dosing requirements and system controls. Testing included collection of data from the online PAA

FIGURE 13.6 Chemical storage and feed facilities final construction (photo courtesy of Kati Bell).

analyzers and grab samples collected at various locations throughout the contact tank. The online PAA analyzers did not perform well in this application and were ultimately discarded. Residual PAA concentrations, associated enterococci concentrations, and other water quality parameters including total suspended solids (TSS), chemical oxygen demand (COD), and total organic carbon (TOC) were used to optimize the operation of the system.

5.0 M.C. STILES WASTEWATER TREATMENT FACILITY

5.1 Background

The M.C. Stiles Wastewater Treatment Facility NPDES permit, issued by TDEC, authorizes the City of Memphis, Tennessee, to discharge treated effluent to the Mississippi River. The facility has a treatment design capacity of 355 m^3/min (135 mgd), with a hydraulic capacity of 657 m^3/min (250 mgd). The facility was originally constructed in 1972 to provide disinfection using gas chlorine; however, issues in the receiving stream with chlorinated disinfection byproducts (DBPs) resulted in TDEC removing the disinfection requirement. It is of note that, at the time of construction, dechlorination was not a common practice. Because of recreational activities on the river, TDEC has added the disinfection requirement to the permit such that the effluent would be required to meet an effluent *E. coli* limit that includes a monthly geometric mean of 126 CFU/100 mL and a daily maximum of 487 CFU/100 mL.

While the least costly method of disinfection would be gas chlorine, the City of Memphis investigated alternative disinfection methods along with reexamining chloramination with dechlorination to determine whether

DBPs could be addressed. High color and low UV transmittance (UVT) in the effluent (Table 13.4), stemming from large industrial dischargers, made UV impractical. Thus, evaluations included chloramination, ozone, ferrate, and PAA. After bench testing and evaluation of the requirements and the costs of these alternatives, the City selected chloramination and PAA for pilot testing at the M.C. Stiles Wastewater Treatment Facility.

5.2 Pilot Testing

Pilot testing was conducted in 2013 to determine and compare disinfection efficacy and the potential formation of DBPs using chloramination/dechlorination and PAA, as well as develop conceptual costs for implementation. Parallel pipe reactors were constructed on-site; sodium hypochlorite and PAA were dosed into the side-by-side reactors (Figure 13.7). Control (undisinfected) samples were collected and analyzed for apparent color, UVT, TSS, *E. coli*, and DBPs; treated samples were analyzed for oxidant residual, DBPs, and *E. coli*.

Comparative capital and operating cost estimates were developed for each technology based on results from pilot testing. Based on technical and water quality results of the pilot testing showing that DBP control was not possible with chloramination/dechlorination, PAA was selected for additional full-scale testing to identify the process control strategy and refine operational costs for meeting disinfection compliance.

5.3 Full-Scale Pilot Testing

In late 2014 and early 2015, a full-scale pilot test was conducted in five phases to identify water quality parameters that correlated well to PAA demand, test feed-forward algorithms based on the two "best-fit" parameters, perform an assessment of the best performing algorithm, and conduct a final verification and qualification of the selected PAA dosing algorithm. Phase 1 was conducted over a 2-week period and showed that color and COD had the best correlations with PAA demand. Phases 2 and 3 were each 2 weeks long and used a PAA dosing algorithm based on color and COD, respectively.

After Phase 3, the color algorithm demonstrated the greatest performance with respect to microbial inactivation and PAA use rates. Phase 4 was a month-long test to assess the disinfection performance of the color-based control algorithm. In Phase 4, adjustments were made to the PAA base dose and the control algorithm. Phase 5 was also conducted over the course of a month and served as the final verification of the algorithm because adjustments were not allowed to be made.

TABLE 13.4 M.C. Stiles Wastewater Treatment Facility effluent water quality characteristics (adapted from CDM Smith, 2014).

	Daily performance data			Draft NPDES limit		
Parameter	Minimum observed	Average or mean[a]	Maximum observed	Daily (minimum; maximum)	Weekly maximum	Monthly maximum
Daily flow, m^3/min (mgd)[b]	150 (58)	247 (94)	610 (232)	355 (135)	—	—
BOD_5, mg/L[c]	5	34	144	86.1	64.7	43.1
TSS, mg/L[d]	1	22	103	104	78	52
pH[d]	6.5	7.2	8.1	6.0–9.0	—	—
E. coli, CFU/100 mL[e]	1.3×10^4	6.0×10^5 (4.4×10^5 as geomean)	1.1×10^7 (1.8×10^6 is 99th percentile)	487	—	126 as geomean
Apparent color, Pt-Co[f]	29	749	2084	—	—	—
True color, Pt-Co[g]	24	619	2000	—	—	—
Apparent UVT, %[h]	0 (1.0)[i]	9.3	36	—	—	—
Filtered UVT, %[j]	0.6 (2.0)[k]	16	71.9	—	—	—

[a]Arithmetic means were reported for all values except *E. coli*, which is also reported as a geometric mean.
[b]Average daily influent flow values; period of record = 9/1/2009–5/31/2013.
[c]Daily values taken from composite effluent samples; period of record = 9/1/2009–5/31/2013.
[d]Daily values taken from and grab effluent samples; period of record = 9/1/2009–12/31/2012.
[e]Daily values taken from and grab effluent samples; period of record = 1/17/2008–7/11/2013.
[f]Daily values taken from composite effluent samples; period of record = 8/11/2011–5/21/2013.
[g]Daily values taken from composite effluent samples; period of record = 8/11/2011–5/21/2013.
[h]Daily values taken from composite effluent samples; period of record = 8/11/2011–5/21/2013.
[i]In parentheses: lowest 10th percentile of apparent UVT data on record as noted in footnote h.
[j]Daily values taken from composite effluent samples; period of record = 8/11/2011–5/21/2013.
[k]In parentheses: lowest 10th percentile of filtered UVT data on record as noted in footnote j.

FIGURE 13.7 Side-by-side pilot reactors for testing chloramination and PAA disinfection (photo courtesy of Kati Bell).

5.4 Procurement

After thorough discussions on various delivery models, the City of Memphis decided to lease the entire PAA system to minimize upfront capital costs. To obtain the most competitive pricing, the City developed a request for proposal (RFP) package to solicit proposals from qualified PAA suppliers. Conceptual design of the PAA storage and feed system was advanced to 60% as these design documents served as the base bid for the RFP. The base bid allowed both 12% and 15% PAA solutions and included a 5-year equipment lease period and options for having the system be maintained and/or operated by the PAA supplier. The City urged PAA suppliers to be creative and allowed bid alternatives for delivery and operation and maintenance of the PAA storage and feed system. Key aspects of the RFP included the following:

- minimum minority/women-owned business enterprise requirements,
- BOD limits,
- feed-forward process control,
- minimum qualifications,
- computational fluid dynamic modeling for mixing verifications,

- multiple PAA injection points,
- alternates for operating the system, and
- maximum residual without quenching.

The City received multiple bids with unique bid alternates to evaluate, allowing them to select the proposal that met their disinfection performance needs while also providing the best value and economic benefit. One proposal included an alternative centering around using 22% PAA, while another included an alternate for a long-term lease (greater than 10 years) centered around a new PAA production facility constructed on property leased from the City at the M.C. Stiles Wastewater Treatment Facility.

The City selected an alternate bid that included a new PAA production facility at the M.C. Stiles Wastewater Treatment Facility, coupled with a smaller PAA storage system at the disinfection tank, operations and maintenance performed by the PAA supplier, and revenue sharing based on the volume of chemical used. The subsequent agreement resulted in a public–private endeavor between the City and the PAA supplier that allowed the City to meet its NPDES requirements while increasing local job creation and tax base.

5.5 Installed Peracetic Acid System

The PAA storage and feed system located adjacent to the disinfection tank consists of the following major components:

- Four 23 100-L (6100-gal) passivated stainless steel storage tanks,
- Multiple online PAA residual analyzers,
- Online color analyzers,
- Effluent flow measurement,
- Programmable logic controller-based control panel,
- Eight peristaltic metering pumps able to deliver a total of 4540 L/hr (1200 gal/hr) of PAA,
- Direct pipeline from the PAA production facility,
- Bulk truck unloading pumps,
- Mixing baffles, and
- Safety equipment.

Construction of the facility began in early 2018 and was completed by August 2018. Figure 13.8 shows the PAA storage and feed system at the M.C. Stiles Wastewater Treatment Facility.

FIGURE 13.8 Peracetic acid storage and feed system at the M.C. Stiles Wastewater Treatment Facility (photo courtesy of Brian Hilts).

5.6 Performance Testing and Operations

Following installation of the PAA storage and feed system, a two-phase performance test was conducted. The first phase included testing the algorithm developed during full-scale PAA pilot testing. The results of the first phase of testing showed that the dosing algorithm based on secondary effluent color did not translate fully to the 2015 full-scale pilot testing, and higher than anticipated PAA doses were required to meet the permitted *E. coli* effluent concentrations. This was attributed to a change in effluent water quality; specifically, undisinfected *E. coli* concentrations were twice as high in 2018 compared to 2015. Adjustments were made and the second phase of the performance test was conducted.

Phase 2 included a month-long test of the system and modified algorithm, where no adjustments were able to be made to the system nor the new algorithm. The PAA system and the dosing algorithm adequately tracked the color changes, which resulted in effluent *E. coli* concentrations and PAA residual values that met the testing requirements with a few notable exceptions. The PAA system, using the modified algorithm, achieved 4 to 5 log reduction in *E. coli* concentration during performance testing, resulting in final disinfected effluent with a geometric mean *E. coli* concentration of 9 MPN/100 mL. On November 10, 2018, the PAA system was placed into full service, making it the largest municipal PAA disinfection system in the world at the time of startup.

Since being placed into operation, the PAA system has had to react to high flows during the first half of 2019 and significant changes in color, all while meeting the PAA residual and effluent *E. coli* limits. The system continues to meet compliance and refinements continue to be made as part of the optimization efforts by the PAA supplier.

6.0 ADAMS FIELD TREATMENT FACILITY

6.1 Background

The Adams Field Treatment Facility (AFTF) is a facility that serves the City of Little Rock, Arkansas. The facility's average day flow is 136 mL/d (36 mgd), and its peak flow is 273 mL/d (72 mgd). The facility, which is the largest WRRF in Little Rock, is managed and operated by Little Rock Water Authority (LRWRA). The AFTF includes primary clarification, activated sludge treatment, and secondary clarification and disinfection. Waste activated sludge is conveyed off-site for treatment and disposal and the Arkansas River receives the discharge of treated effluent from the facility.

A stringent total residual chlorine (TRC) effluent discharge limit, and the inherent safety risks associated with the storage of hazardous chlorine gas, prompted LRWRA to replace chlorine disinfection with UV disinfection in 2008. A new low-pressure, open-channel UV system was constructed adjacent to the chlorine disinfection building and the interior of the chlorine building was retrofitted for housing UV power supply centers and system controls.

The UV system is arranged in four parallel open channels to achieve seasonally variable disinfection limits set forth in the NPDES permit (Table 13.5). Each channel contains 120 lamps per bank, with lamps oriented parallel to flow to deliver a minimum germicidal UV dose of 24.8 mJ/cm^2 under the following conditions:

- Secondary clarifier effluent fecal coliform $< 10^5$ CFU/100 mL,
- Clarifier effluent TSS < 30 mg/L,

TABLE 13.5 Relevant effluent discharge limits for Adams Field Treatment Facility.

Effluent characteristics	Discharge limitations		Monitoring requirements	
	Monthly avg.	7-day avg.	Frequency	Sample type
Fecal coliform May–September	200 CFU/100 mL	400 CFU/100 mL	Two/week	Grab
Fecal coliform October–April	1000 CFU/100 mL	2000 CFU/100 mL	Two/week	Grab

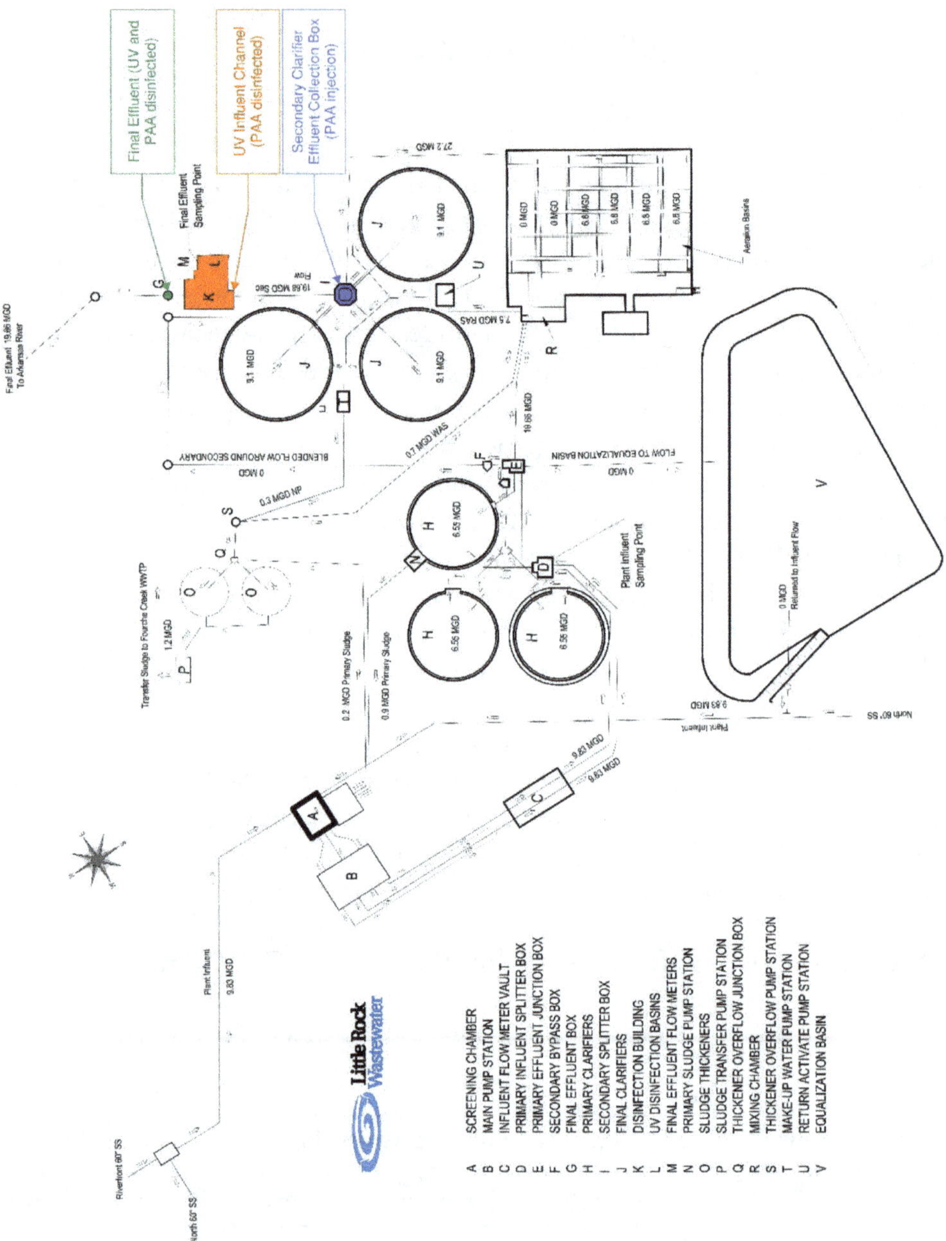

FIGURE 13.9 Schematic of flow through Adams Field Treatment Facility, with the location of PAA injection identified (reprinted with permission from the Little Rock Water Reclamation Authority).

- Clarifier effluent UVT ≥ 65%, and
- UV influent flow of 273 mL/d (72 mgd) (or less).

6.2 Driver for Implementation of Peracetic Acid

Since installation of the UV disinfection system in 2008, Adams Field has experienced low UV transmittance, or UVT, (less than 65%) in secondary effluent −50% of the time. The selected UV equipment for this facility was designed to meet effluent discharge limits for fecal coliform when UVT is greater than or equal to 65%.

In 2014, on-site testing was used to assess the performance of the UV system. Results of this testing showed that, in addition to low UVT, high TSS during peak flows impaired UV dose delivery. As part of that evaluation, alternative, supplemental disinfection strategies were reviewed to either replace or augment the performance of the existing UV system. The outcome of the evaluation was that PAA should be evaluated as a standby chemical disinfectant at AFTF because of its following potential benefits:

- PAA can be fed as needed (when facility UVT falls below 65%),
- PAA can be added upstream of UV disinfection, and
- PAA is unlikely to require residual destruction before discharge to the receiving stream.

Peracetic acid (15% product) was successfully pilot tested at the AFTF and LRWRA installed permanent PAA chemical storage and feed facilities in the fall of 2016 so that they could apply PAA to secondary clarifier effluent to enhance UV disinfection. Figure 13.9 presents a schematic of treated flow through the AFTF and the location of supplemental PAA injection.

6.3 Permitting and Implementation of Supplemental Peracetic Acid Disinfection

Careful consideration was given to the potential impact of residual PAA on the receiving stream and on required quarterly WET compliance. As part of the pilot tests, LRWRA performed modified WET tests to quantify the no-effect levels of residual PAA in AFTF effluent samples. Results led the permitting authority (Arkansas Department of Environmental Quality) to grant AFTF a temporary permit modification for discharge of effluent containing up to 2 mg/L of residual PAA (Bailey, 2016), with the following additional requirements:

- Daily monitoring (grab sample) of effluent fecal coliform must be conducted when PAA is applied.

TABLE 13.6 Disinfection performance while applying supplemental PAA at Adams Field Treatment Facility (August 2016 to March 2018).

	PAA dose at final clarifier (mg/L)	Effluent PAA residual (mg/L)	Final clarifier effluent, fecal coliform (CFU/100 mL)	UV influent, fecal coliform (CFU/100 mL)	Effluent, fecal coliform (CFU/100 mL)
Max.[a]	1.55	3.46	415 000	177 000	1100
Mean[b]	0.89	0.82	78 000	16 000	157
Min.[c]	0.48	0.31	3900	63	2

[a]Maximum of all samples collected
[b]Mean of all samples collected
[c]Minimum of all samples collected

- Continuous use of UV disinfection must be collected, even when PAA is applied.
- A 24-hour composite effluent sample must be collected for quarterly WET testing on a day when PAA is applied (for a minimum of one WET with PAA residuals per year).
- Daily reporting of effluent PAA residuals (grab samples) must be collected when PAA is applied.
- Daily records must be kept of PAA application rates and duration of feed.

6.4 Supplemental Peracetic Acid System Performance

As shown in Figure 13.9, LRWRA applies PAA upstream of the UV disinfection system; the point of application provides less than 4 minutes of contact time from the clarifier effluent collection box and through a 1.8-m (72-in.) reinforced concrete conduit that links the secondary clarifiers to the UV disinfection facility. Even with this short contact time, a dose of <2 mg/L PAA provides >2-log fecal coliform inactivation, as demonstrated in Table 13.6, which also provides a summary of final effluent fecal coliform performance following UV disinfection.

6.5 Lessons Learned from Supplemental Peracetic Acid

The success of this project, which includes improved disinfection compliance as well as improvements in frequency of UV maintenance and reduced UV power consumption, is attributable to the contributions of LRWRA facility operations staff. Specifically, LRWRA staff assess the need and potential benefits of PAA application, given anticipated and observed influent flow and loadings each day, and provide operational intelligence to optimize use of the standby PAA system.

7.0 REFERENCES

Bailey, J. (2016). *Full scale trial of disinfection using peracetic acid at Adams Field Facility* (edited by Arkansas Department of Environmental Quality). North Little Rock, AR: Office of Water Quality.

CDM Smith. (2013). Recommended disinfection design doses for PAA and chlorine. A technical memorandum prepared for the Gulf Coast Authority.

CDM Smith. (2014). Stiles WWTP preliminary engineering report. Prepared for the City of Memphis, TN.

CDM Smith. (2015). Peracetic acid pilot results, summary and disinfection comparison. Prepared for Metro Vancouver under NORTHWEST LANGLEY WWTP RFQ NO. LB1407, June 2015.

Environment Canada. (2008). *Procedure for pH stabilization during the testing of acute lethality of wastewater effluent to rainbow trout* (Report EPS 1/RM/50). Ottawa, Ontario, Canada: Environment Canada, Method Development and Applications Section, Environmental Science and Technology Centre, Science and Technology Branch.

Kitis, M. (2003) Disinfection of wastewater with peracetic acid: A review. *Environment International,* 30, 47–55.

Lapczynski, K. (2016). TUB's wastewater treatment plant goes chlorine-free. *The News*, Tullahoma Tennessee. Retrieved from https://www.tullahomanews.com/news/local/tub-s-wastewater-treatment-plant-goes-chlorine-free/article_6c368fe8-040c-5bf6-8fc1-ed6c90210347.html

Monarca S., Feretti D., Zerbini I., Zani C., Alberti A., Richardson S.D., et al. (2001). Studies on mutagenicity and disinfection by-products in river drinking water disinfected with peracetic acid or sodium hypochlorite. Proceedings of IWA World Conference, Berlin, Germany.

Monarca S., Richardson S.D., Feretti D., Grottolo M., Thruston A.D., Zani C., et al. (2002). Mutagenicity and disinfection by-products in surface drinking water disinfected with peracetic acid. *Environmental Toxicology and Chemistry,* 21, 309–318.

Young, S. (2016). Conversion to the peracetic acid disinfectant for the Tullahoma Wastewater Treatment Plant—Revised engineering report, submitted to TDEC. Retrieved from http://tdec.tn.gov:8080/pls/enf_reports/f?p=9034:34051:::NO:34051:P34051_PERMIT_NUMBER:TN0023469

14

Peracetic Acid—A Look Forward

Philip Block & Kati Bell

1.0 ADVANCING PERACETIC ACID IN WASTEWATER DISINFECTION

Peracetic acid (PAA) has been used commercially as a wastewater disinfectant for more than 20 years in Europe, 10 years in the United States, and six years in Canada. As a result, peracetic acid is in the process of transforming from an "innovative and emerging" disinfection technology to an established method in the toolbox of the wastewater professional. At the time of this manual, the path to maturity is still a long one, because, in the United States, less than half of the regulatory agencies have granted approval for use because of a regional lack of familiarity with PAA, the perceived risks of approving new disinfection technologies, state-specific requirements of adopting aquatic toxicity residual endpoints, and the need to expand the database on PAA's efficacy on viruses of human concern. Until the point where PAA is as broadly accepted as chlorine, ultraviolet

(UV), and ozone disinfection are, the growth of PAA use will be governed by the regional regulator and wastewater utilities in need of an alternative technology to meet their specific needs.

Although PAA has been approved for registration as a wastewater disinfectant with allowable PAA residuals by countries such as the U.S. and Canada, there are no national water quality criteria for PAA in either of these countries. As a result, each state, province, and/or region is developing their own water quality criteria, often with little commonality in test methodology or acceptance criteria. Often, development of such criteria does not even begin unless a local treatment facility triggers the assessment by submitting a modification to their permit to allow PAA use as their disinfection technology. Data gaps critical for analysis might include approved aquatic toxicity testing via a specific protocol for that state or analysis of toxicity impacts on local fish and other aquatic species. In addition, PAA efficacy data on specific bacterial, and, more particularly, viral pathogens specific to state or regional requirement may need to be generated. A similar situation exists in Canada, where the Pest Management Regulatory Agency of Health Canada has approved peracetic acid for use within the country, but each province retains local approval of permitted residuals. As a result, British Columbia has led the way in terms of approving PAA for use, while other provinces lag in their acceptance of PAA. The situation in Europe is likewise similar, with approval controlled by each country and local region. While the Nosedo wastewater treatment plant in Milan, Italy, was not only one of the first adopters of PAA as a wastewater disinfectant, and one of the largest users of PAA by volume, expansion of PAA throughout the continent has been very limited.

Although PAA test methods are equivalent to the chlorine colorimetric methods currently in use, there is currently no U.S. EPA standard test method for the measurement of PAA in wastewater, nor is there likely to be one in the foreseeable future. Such lack of a "standard" U.S. EPA approved PAA analytical method has been a hurdle to PAA approval for use in several states of the United States. However, a PAA *Standard Method* (Standard Method, 2019) recently was published in 2019 through the work of a joint task group, which included members from the U.S. EPA, wastewater plant operators, industry members, and method developers

1.1 Innovations in Peracetic Acid Disinfection

1.1.1 Use of Peracetic Acid in Combination with Ultraviolet Disinfection Systems

Combining PAA and ultraviolet (UV) disinfection provides the opportunity to take advantage of the unique disinfection capability each technology has with respect to microbial kill, perhaps in a synergistic manner, while counterbalancing their limitations. The pairing of these two disinfection technologies has the potential to increase performance and enhance economic efficiency, compared to deploying each independently. While UV disinfection has had wide acceptance for wastewater treatment, its application can be limited in wastewaters with high particulate concentration, such as suspended solids, and in water with low UV transmittance. Peracetic acid may be limited in waters with high oxidant demand and for microbial species that require high dose concentrations. Because of their different modes of action, combining PAA with UV may overcome situations where one or both may be inappropriate for use independently.

Peracetic acid may be deployed in conjunction UV for several reasons. If a synergistic effect is demonstrated, where the combination of PAA and UV exceed the performance of the additive effect of using each method separately, then an economic savings may be realized because of reduction in the required UV footprint (capital expense), UV power usage (operational expense), or PAA usage (operational expense). Peracetic acid can be used for "peak-shaving" UV systems so that the UV system may be designed for average flows, with the addition of PAA used for periods of high flows, such as rain events.

For UV systems already in place, the addition of PAA may allow aging systems to meet permit requirements without significant redesign or construction, or it may help operational UV systems meet new regulatory requirement, such as new target organisms or low target organism concentrations, without the need for expanding the UV footprint. Finally, addition of PAA may help reduce the maintenance cycle of UV systems by maintaining bulb performance because of reduction in algae and mineralization buildup on the bulb.

The combination of PAA and UV disinfection was investigated (Caretti & Lubello, 2003) for several microorganisms in a wastewater passed through a pilot reactor. It was demonstrated that higher disinfection efficiency was observed for the combined technology for all species tested (total coliforms, fecal coliforms, fecal *streptococcus, Escherichia coli,* and *Pseudomonas aeruginosa*). The results of the study by Chen, Dong, and Gehr (2005) for fecal coliforms in wastewater demonstrated, at a minimum, an additive benefit of the PAA and UV and a statistically significant interaction between the PAA and UV, but failed to demonstrate a true synergistic effect.

A study (Koivunen & Heinonen-Tanski, 2005) in peptone water demonstrated true synergy of the PAA/UV system for *E. faecalis*, *E. coli,* and *Salmonella enteritidis* and, for several treatment scenarios, coliphage MS2. De Souza et al. (2015) were not able to demonstrate synergy for all species and test ranges but showed high efficacy of microbial inactivation for *E. coli*, total coliforms, and coliphages. In contrast, no synergistic benefit was observed (Weng et al., 2018) under common water resource recovery facility (WRRF) conditions (0.5 to 2 mg PAA/L and 20 mJ/cm^2 UV dose) in a wastewater matrix for the reduction of MS2 bacteriophage or murine norovirus populations, and it was suggested this was attributable to the presence of radical scavengers in the municipal wastewater.

One of the potential benefits of using PAA in conjunction with UV is the ability to use less power for the UV disinfection to achieve equivalent microbial reduction performance with UV alone. This was demonstrated (New York State Pollution Prevention Institute, 2013) using a bench-scale collimated beam apparatus. The study was run with wastewater, and the samples were enumerated for fecal coliforms and enterococci. Peracetic acid was added to the wastewater before UV radiation. The results from this study for fecal coliforms show the addition of PAA enhanced the UV treatment performance, allowing the equivalent coliform log reduction to be achieved at a lower UV dose across a wide range of operating doses, as shown in Figure 14.1.

Similar results were observed for enteroccoci at UV doses of 10 mJ/cm2, but this was not observed at higher UV doses of 20 and 40 mJ/cm2. Martin et al. (2007) reported similar outcomes for fecal coliforms, where a

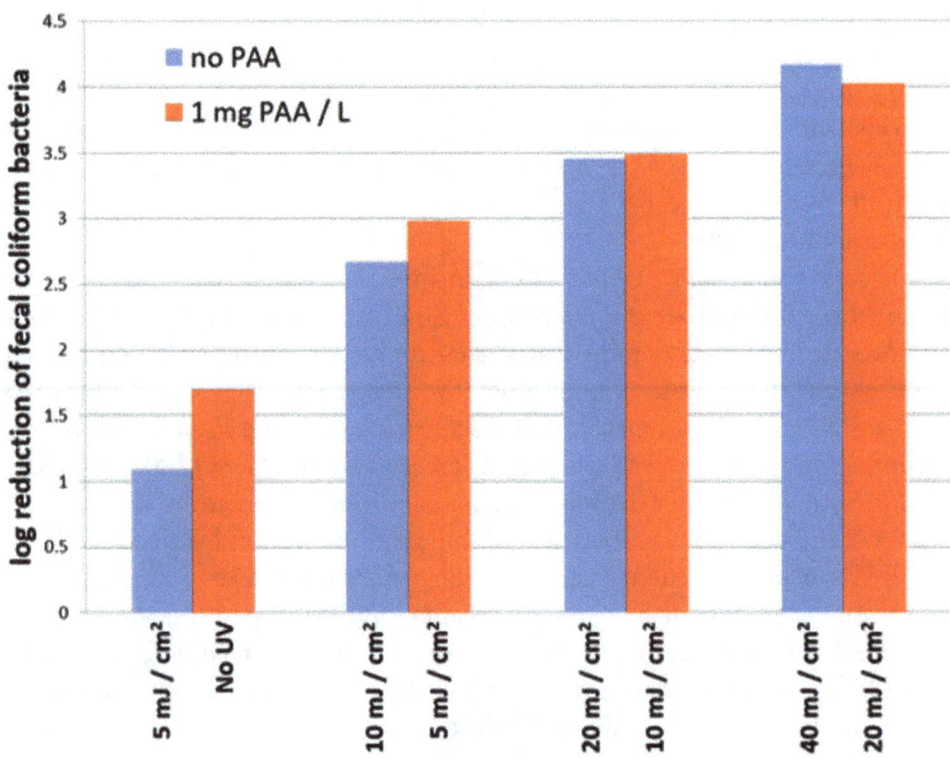

FIGURE 14.1 Efficacy of PAA and UV combinations on fecal coliform reduction.

UV dose of 40 mJ/cm2 yielded 4-log inactivation, similar to 2-mg PAA/L combined UV at 20 mJ/cm2. The combination of PAA and UV is now being applied at several wastewater systems, including at facilities in New Jersey, Alabama, and Kentucky (PeroxyChem, 2019).

1.1.2 Wet Weather Applications

Combined sewer overflow (CSO) systems combine wastewater with stormwater in the same sewer, typically when the rainfall event exceeds the design capacity of the sewer system and the effluent needs to be discharged to surface waters. They are typically defined by low retention times and high organic loadings, which is more reminiscent of primary treated wastewaters. Wet weather events are often infrequent (several times a year) and may result in high water flowrates. Several disinfection technologies are used for CSO treatment, including chlorination, UV, and chlorine dioxide. Chlorine is the most widely used disinfectant for CSO applications (Chhetri, Thornberg, Berner, Gramstad, Ojstedt, Sharma, & Andersen, 2014), but suffers from the formation of chlorinated organics because of high organic loadings of the incoming rainwaters, high sensitivity of chlorination efficiency with pH variations (Chhetri et al., 2014), formation of chloramines with nitrogen compounds in the raw water, and the need for long-term, on-site storage of the chemical, which decomposes somewhat rapidly over time.

In contrast, PAA does not generate disinfection byproducts in the presence of organic demand in the raw water and decomposes rapidly in the environment. Thus, use of PAA for CSO or wet weather events may not require a quenching step before discharge that is often needed for chlorination processes.

Storage of a liquid disinfectant can be a significant design consideration when dealing with CSO or wet weather events. Sporadic use and potentially high use rates may require large volumes of disinfectant to be stored, often at remote locations. Peracetic acid has a much longer storage life than hypochlorite, with the stability of PAA lasting up to 1 year, depending on the solution temperature. As a result, bulk PAA solutions do not need continual analysis to ensure they are maintaining their solution concentration and, therefore, proper dose rates are more readily ensured.

However, challenges remain when considering the use of PAA for intermittent, extended storage applications such as wet weather events. Supply chain requirements to deliver sufficient quantities of PAA to a potentially remote site must be accounted for, and enough storage capacity to meet multiple rain events must be designed into the disinfection system to ensure an adequate supply of PAA is on hand to meet the most conservative rainfall estimates. If bulk storage is warranted, remote tank gauging can be used to ensure proper PAA tank levels are maintained, and a vendor-managed delivery system can be used to reduce the risk of product shortages. Akin to hypochlorite, PAA is a strong oxidant, and adequate safety precautions such as security fencing must be installed to ensure product integrity, especially in remote sites.

U.S. EPA has listed PAA as an alternative disinfection technology for CSOs. Laboratory investigation has demonstrated that CSO-like waters containing suspended solids demonstrated that PAA was more efficient than hypochlorite for all solids size fractions, and that particle size had little effect on PAA disinfection efficacy (McFadden et al., 2017). Combining pretreatment processes, such as chemical coagulation and particle separation, was shown to have a beneficial effect on PAA disinfection efficiency (Chhetri, Bonnerup, & Andersen, 2016). Effective PAA dose concentrations for CSOs typically will be higher than the equivalent concentrations needed for secondary wastewater treatment because of higher PAA demand from organic and solid loading in the CSO effluent.

Laboratory jar testing can be adequate to determine the appropriate PAA concentration needed to meet the bacterial reduction requirements under the site-specific contact times. However, obtaining a suitable CSO sample during a rain event can be problematic because forecasting the occurrence of such an event is difficult. Primary effluent from the WRRF can be used as a substitute, but often will overestimate the amount of PAA needed. If laboratory time is flexible, it is best to collect a CSO for analysis during a typical rain event to best assess the PAA dose requirements.

1.1.3 Biosolids Disinfection

Given PAA's demonstrated disinfecting capabilities for wastewater effluents, it has the potential to provide similar disinfecting potential for wastewater sludge or biosolids. In addition, PAA may serve as a preconditioning agent before dewatering to improve dewatering process efficiency and, in turn, reduce transportation and disposal costs (Zhang, Cao, Wang, Ma, Xia, & Dehong, 2016). Recent bench-scale work (PeroxyChem, 2015) on sludge from a WRRF in northern Florida has shown that PAA can disinfect sludges with dose concentrations of 500 mg PAA/L and contact times of 60 minutes or greater, with significant reductions in fecal coliforms, *E. coli,* and Salmonella (Table 14.1). Enteric viruses and Helmuth egg reductions were not monitored. During this test, no reduction in total volatile organics were observed.

While PAA may be suitable as a biosolids disinfection technology, the economics of applying a large dose of PAA must be compared to alternative sludge disinfection methods, and demonstration on enteric viruses

TABLE 14.1 Reduction of indicator organisms in wastewater sludge (PeroxyChem, 2015).

PAA dose	Dry weight	*E. Coli*			Fecal coliforms			Salmonella		
mg/L	%	CFU/g	Log CFU/g	Log CFU/g reduction	CFU/g	Log CFU/g	Log CFU/g reduction	MPN/kg	Log MPN/kg	Log MPN/kg reduction
0	2.3	1.40 E+06	6.15	0.00	2.6 E+06	6.42	0.00	279.8	2.45	0
500	2.3	43.5	1.64	4.51	1.5 E+03	3.18	3.24	1.1	0.04	2.41
750	2.4	41.7	1.62	4.53	6.3 E+02	2.80	3.62	1.1	0.04	2.41
1000	2.4	41.7	1.62	4.53	4.2 E+01	1.62	4.80	1.1	0.03	2.43

and Helmuth eggs must be performed, before commercialization of this technology.

Peracetic acid use as a biosolids preconditioning agent for dewatering purposes was demonstrated at the laboratory scale (Su et al., 2004; Sun et al., 2011; Zhang et al., 2016). Peracetic acid was shown to dissolve extracellular polymeric substances and oxidizing cells, promoting release of polymers and reducing sludge volume (Su et al., 2004). This was enhanced by the presence of ultrasonic waves. It was also observed that PAA treatment increased biodegradability of the organic compounds present (Sun et al., 2011) and that microwave radiation showed significant synergistic effects with PAA. Peracetic acid can effectively solubilize sludge particles, especially for protein-like molecules and polysaccharides; post-PAA treatment, sludge floc structure can be reconstituted via the use of inorganic coagulants (Zhang et al., 2016). Appels et al. (2011) demonstrated that PAA effectively leads to solubilization of organic materials in waste activated sludge at dose concentrations of 25 g PAA/kg dry solids. In addition, at these PAA concentrations, a 21% increase in biogas production was observed post-anaerobic digestion.

1.2 Advanced Research into Novel Peracetic Acid Applications

As PAA becomes more widely used as a wastewater disinfectant, additional research into novel PAA application uses as a disinfectant or pre-treatment chemistry throughout the wastewater treatment process continues. The following subsections present a few examples of these areas of research.

1.2.1 Enhanced Microbial Control

Improving the performance of PAA on viruses and other pathogens of concern has been the subject of ongoing research. Typically, this involves combining PAA with other materials to generate hydroxyl and/or peracetyl radicals, which are thermodynamically and kinetically capable of reacting with more refractory compounds. One example, as discussed in Chapter 6, is the combination of PAA with UV irradiation. Studies have also investigated the use of adding transition metals such as silver and copper (Velasquez et al., 2008) to PAA for improved microbial efficacy.

Velasquez et al. (2008) demonstrated a synergistic effect on PAA efficacy when combined with silver in disinfecting primary treatment effluent, reducing the contact time in half to reach equivalent log reductions in fecal coliforms and pathogenic bacteria (*Vibrio cholerae, Shigella* spp, *and P. aeruginosa)* and well-significant reductions in helminth eggs. Researchers are currently investigating combinations of PAA with other disinfectants and compounds to look for synergistic benefits.

1.2.2 Reduction of Contaminants of Concern

As described in Chapter 6, PAA is being investigated as an oxidative treatment to remove contaminants of concern. These include pharmaceuticals, endocrine disruptors, and disinfection byproduct precursors. Peracetic acid oxidation of such contaminants may lend itself not only to wastewater applications, but to a multi-barrier approach to water reuse and as a "pre-oxidant" in drinking water treatment. As part of an advanced oxidation system, PAA may be used via direct oxidation or in combination with UV or transition metal catalysis.

1.2.3 Membrane Cleaning

Peracetic acid is listed by the National Sanitation Foundation as a membrane cleaner for water treatment. It has been used in the medical and food industries for controlling biofilm formation and has been considered for use in reverse osmosis facilities. However, its use in membrane disinfection is limited because of potential oxidative effects on polymeric membranes (Nguyen, Roddick, & Fan, 2012). This may increase the permeability of the membrane and reduce its effectiveness and lifetime. Still, PAA may be efficient for disinfecting the pretreatment section components in membrane filtration processes such as pipes, manifolds and other hard-to-reach stagnant flow areas (Saad, 1992). Further research is currently underway investigating the potential of PAA to enhance membrane performance and longevity.

While conceptual, incorporation of PAA to a membrane bioreactor (MBR) may provide the MBR with an additional carbon source as a nutrient from the residual acetic acid from PAA while helping to maintain control of biofilm and bio-floc formation on the filter. However, proper control of PAA dosing would be a must so as not to disrupt the activated sludge bioreactivity.

2.0 PERACETIC ACID DISINFECTION FOR EMERGING MICROBIAL CONTAMINANTS

The National Institute of Allergy and Infectious Diseases defines "emerging infectious diseases/pathogens" as those "that have newly appeared in a population or have existed but are rapidly increasing in incidence or geographic range." Many of these emerging pathogens of concern are pathogenic viruses and pose serious threats to public health. Newly emerging viruses include the Ebola virus, severe acute respiratory syndrome (SARS), Middle East respiratory syndrome (MERS) coronavirus, and the avian influenza virus. The 2009 swine flu pandemic reminded us that the Spanish flu killed over 40 million people.

Because the occurrence of emerging viral pathogens is less common and less predictable than established pathogens, few if any U.S. EPA-registered disinfectant product labels specify use against these types of infectious agents. In 2016 U.S. EPA provided a voluntary, two-stage process to enable use of certain U.S. EPA-registered disinfectant products against emerging viral pathogens not identified on the product label. Through this process, a company can apply for an emerging viral pathogen claim, even before an outbreak occurs, based on previous U.S. EPA-approved claims for harder-to-kill viruses. The U.S. EPA approval process allows a company to make off-label claims for disinfectants in the in the event of an outbreak such as the recent severe acute respiratory syndrome-coronavirus 2 (SARS-CoV-2), the causative agent for COVID-19 (coronavirus disease 2019), first identified in Wuhan China.

U.S. EPA's emerging viral pathogen guidance was triggered for PAA disinfection of SARS-CoV-2 on January 29, 2020 when U.S. EPA allowed emergency use of PAA products to reduce SARS-CoV-2 concentrations based on use rates for norovirus. While the emergency uses were specified for applications other than wastewater disinfection, PAA products have been provided to various municipalities for sanitization procedures used for continued operations of essential services, including wastewater treatment, during the pandemic.

3.0 CONCLUSION

Peracetic acid is a potent disinfection tool for the wastewater professional. Its advantages as a wastewater disinfectant allow the design engineer to meet regulatory and socioeconomic criteria for a given project that may have been difficult to meet with other comparable technologies. Peracetic acid's biocidal efficacy, lack of disinfection byproduct formation, low environmental impact that does not require quenching agents, positive safety profile, long shelf life, and relatively low capital cost make it a viable option for many WRRFs.

While PAA is not universally approved for implementation at this time its continued expansion to multiple, and larger facilities and its growing number of years of performance results will usher in its transition to a mature and reliable technology over the coming years. Today's PAA production capacity and supply chain capability ensures that PAA availability and delivery can match the needs of tomorrow's future growth demands.

From its beginnings as a medical biocide replacement for glutaraldehyde more than 30 years ago to its recognized performance advantages in the food safety marketplace, PAA today is being established as a viable technology within the wastewater disinfection space. Looking to the future, as research into the its performance benefits continues, PAA will play a role in wet weather disinfection, advanced oxidation systems, and raw water conditioning to remove contaminants of concern that can impact human health. Additionally, PAA will continue to play a role in disinfection of emerging microbial pathogens both as an effluent disinfectant, as well as a sanitization agent for maintaining operations during future emerging pathogen epidemics and pandemics.

4.0 REFERENCES

Appels, L., van Assche, A., Willems, K., Degreve, J., van Impe, J., & Dewil, R. (2011). Peracetic acid oxidation as an alternative pre-treatment for the anaerobic digestion of waste activated sludge. *Bioresource Technology,* 102, 4124–4130.

Caretti, C., & Lubello, C. (2003). Wastewater disinfection with PAA and UV combined treatment: A pilot plant study. *Water Research,* 37, 2365–2371.

Chen, D., Dong, S., & Gehr, R. (2005). Alternative disinfection mechanisms for wastewaters using combined PAA/UV process. *Disinfection* 2005, Water Environment Federation, 886–906.

Chhetri, R., Bonnerup, A., & Andersen, H. (2016). Combined sewer overflow pretreatment with chemical coagulations and a particle settler for improved peracetic acid disinfection. *Journal of Industrial and Engineering Chemistry,* 37, 372–379.

Chhetri, R., Thornberg, D., Berner, J., Gramstad, R., Ojstedt, U., Sharma, A., & Andersen, H. (2014). Chemical disinfection of combined sewer overflow waters using performic or peracetic acids. *Science of the Total Environment,* 490, 1065–1072.

De Souza, J., Valdez, F., Jeranoski, R., Vidal, C., & Cavalinni, G. (2015). Water and wastewater disinfection with peracetic acid and UV radiation and using advanced oxidative process PAA/UV. *International Journal of Photoenergy*, 1–7.

Koivunen, J., & Heinonen-Tanski, H. (2005). Inactivation of enteric microorganisms with chemical disinfectants, UV irradiation and combined chemical/UV treatments. *Water Resources*, 39, 1519–1526.

McFadden, M., Loconsole, J., Shockling, A., Nerenberg, R., & Pavissich, J. (2017). Comparing peracetic acid and hypochlorite for disinfection of combined sewer overflows: Effects of suspended solids and pH. *Science of the Total Environment*, 599, 533–539.

New York State Pollution Prevention Institute. (2013). VigorOx WWT II Performance Evaluation, report to PeroxyChem.

Nguyen, T., Roddick, F., & Fan, L. (2012). Biofouling of water treatment membranes: A review of the underlying causes, monitoring techniques and control measures. *Membranes*, 2, 804–840.

PeroxyChem. (2015). Sludge disinfection study (unpublished). Philadelphia, PA: PeroxyChem.

PeroxyChem. (2019). Internal commercial and installation data. Philadelphia, PA: PeroxyChem.

Saad, M. A. (1992). Biofouling prevention in RO polymeric membrane systems. *Desalination*, *88*(1–3), 85–105.

Standard methods for the examination of water and wastewater. (2019). Method 4500-PAA. https://www.standardmethods.org/doi/abs/10.2105/SMWW.2882.220

Su, F., Xing, X., & Sun, X. (2004). Reduction of excess sludge by peracetic acid. *Journal of Chemical Engineering, CHN University*, 18, 471–476.

Sun, D., Guo, S., Ma, N., Du, Y., Dong, X., Zhang, X., Xue, M., & Ma, C. (2011). Pretreatment of excess sludge by microwave combined with peracetic acid. *Environmental Pollution Control*, 33, 50–53.

Velasquez, M., Yanes-Noguez, I., Himenez-Cisneros, B., & Pabello, V. (2008). Adding silver and copper to hydrogen peroxide and peracetic acid in the disinfection of an advanced primary treatment effluent. *Environmental Technology*, 29, 1209–1217.

Weng, S., Dunkin, N., Schwab, K., McQuarrie, J., Bell, K., & Jacangelo, J. (2018). Infectivity reduction efficacy of UV irradiation and peracetic acid—UV combined treatment on MS2 bacteriophage and murine norovirus in secondary wastewater effluent. *Journal of Environmental Management*, 221, 1–9.

Zhang, W., Cao, B., Wang, D., Ma, T., Xia, H, & Dehong, U. (2016). Influence of wastewater sludge treatment using combined peroxyacetic acid oxidation and inorganic coagulants re-flocculation on characteristics of extracellular polymeric substances (EPS). *Water Research*, 88, 729–739.

Index

CPSIA information can be obtained
at www.ICGtesting.com
Printed in the USA
BVHW012121150121
597993BV00008B/45